BLACKDEATH 23

My Journal as an Army Helicopter Pilot in Iraq

by ROBERT MILLS
Chief Warrant Officer 2
United States Cavalry

BLACKDEATH 23
My Journal as an Army Helicopter Pilot in Iraq
by ROBERT MILLS

Copyright © 2014

Additional copies of this book are available from the author:
Robert Mills
WWW.BLACKDEATH23.COM

Printed in the United States of America
Wise Publications
Custom Book Manufacturing
809 East Napoleon Street
Sulphur, LA 70663
337-527-8308
wisepublications@yahoo.com
www.wisepublications.biz

Acknowledgments

I feel honored in sharing my experiences with everyone. To my wife Roberta and daughters, Naelyn and Madison, I apologize for missing a portion of your life while I was away. You were strong when I was not home to be strong for you. To my parents Bud and Linda, thank you for planting me on solid ground at a young age. To my brother Rod, thank you for planting the seed of possibility in me. To Bob and Marilyn Stevens, thanks for being there for the girls when I wasn't able to. To my best friend Jason Moore, thank you for being my second brother and spending countless hours drawing helicopters, airplanes, and aerial combat formations and dreaming that one day this may be possible. Doctor Jack Rushin (Duke), I'm not sure I would have made it through Basic Training if you hadn't continually kept adjusting me during my pre-entry workouts. Thanks for keeping me straight and supporting the military. To Greg and Angie West, thank you for your friendship in tough times and your unending support for the military and the United States.

Deployments are difficult, particularly when you're living and operating on bare essentials. Our Troop purchased an internet system during OIF 07-09 by combining funds and through donations. Thank you Ival Gregory for your strong financial support that allowed us to stay in contact with our families back home. Your donation was a blessing.

To all the Cavalry Troopers I served with, I say God Bless the Cav! I love you all and was proud to serve with you. The Troopers of Blackdeath will always hold a special place in my heart and mind. Thank you for serving beside me.

To my editors, Kerri Ponder and Carrie Smothers, thank you. I greatly appreciate your work on this.

To the Lord God, thank you for calling me to this mission and protecting me.

Yea, though I walk through the valley of the shadow of death,
I will fear no evil; for thou art with me.
Psalms 23:4

Prologue

I grew up in a small town in southeast Missouri with a passion for the outdoors. I spent the majority of my time away from work, hunting and fishing. I met my wife Roberta in my home town and we married two years later in 1993. By 1996 we had two daughters, Naelyn and Madison, and I was well on my way to a career in electronics and computer repair.

In 1999 I moved into a Computer Analyst position for a large corporation. We were very busy with the Y2K transition issues that everyone feared would halt all computerized processes. Corporations were pouring money into technology in hopes that they would not be bitten by the Y2K bug. Late in the year 2000, after the world didn't shutdown and corporations saw that everything was working as planned, they quickly cut funding that had previously been spent on information technology. Although understaffed, my company identified my position as one that was vulnerable. Fearing the vulnerability and being fed-up with the situation, I began praying to God. "Where do you want me Lord?"

As the months passed, I felt that I needed to make a drastic change but wasn't sure what it was. I shared with my wife what I felt God was telling me. I think it made her nervous not knowing what was going to happen. I'm sure I was as nervous as she was. I couldn't get clarity on the answer so shared it with my brother Rod. He was, and currently is, a Chaplain in the Army. His reply was, "Why don't you do what you've always said you want to do, fly helicopters?" As a child I would spend hours drawing helicopters, strategizing with my best friend Jason on hangars and attack planning but this is real life. I rarely even saw a helicopter much less thought of flying one. Who has the money for lessons? It costs about $50-60K to learn how to fly. I was poor, making just enough to feed the kids it seemed like. He said the Army has a program called "High School to Flight School." You should check it out.

I thought about that for several days and couldn't get it out of my mind. It couldn't hurt to check it out so I went to the recruiter's office, introduced myself, and asked him to tell me about the program. He didn't know what I was talking about. As far as he knew, there was no such thing. He went on to offer me the option to sign up for the Army and then work my way to a flight slot. I said nope, not interested. I told him to check into the program and call me when he figured it out.

He called me the next day and said, "What do you know, the program does exist!" Imagine that. Now that I knew this was an option, I went back to visit with him some more. He showed me some of the helo pictures and an intro video for Aviation Special Operations that were very enticing. I liked what I saw but there was a mountain of paperwork that I need to do to be accepted. After praying about it some more, I made the decision to go through the preliminary paperwork and physicals.

When I came home and told my wife what I wanted to do, she thought I was crazy. She starred at me for a long time with a very puzzled look on her face. Once we made it through the shock, she asked are you sure? Yep, I think this is the change God has for us. Having suffered a fall in the past year from 22 feet, the military wanted me to under-go a multitude of tests, some at my own expense, one of which was a CT-Scan. This would cost $1000! I didn't have that kind of money. A CT-Scan had been completed after my fall so they agreed to accept the results of that. Additionally, I was required to pass a flight physical conducted by a flight doctor at Fort Campbell, KY. This process took several months to complete but by May 2001, I passed all of their requirements. The last step was to be interviewed by a panel of Army Officers in St. Louis at the Entry Processing Center. Myself and two other civilian candidates waited patiently for our interviews. After an hour or so they called me in. I walked into the room and there was a long table with five Army Officers ranging from Captains to Majors sitting behind the table and facing an open chair in the middle of the room, my seat. It looked cold and not so inviting. After sitting down, they asked me several questions about myself, my job history, my family, and why I thought I could be an Army combat helicopter pilot. I responded, "It's something I've always dreamed of doing and I've never failed at anything I've set my mind to."

It took four weeks for the board results to come back. My recruiter called me and gave me the good news, "You've been accepted!" What a great feeling. The entry process took nine months but I was finally accepted and could fulfill a childhood dream. This news came shortly after my dad was diagnosed with colon cancer. His surgery was scheduled to be during my Basic Training class so I delayed my entry date to be with him. Unfortunately, they delayed it a second time. I had already delayed leaving for Basic once so he didn't want me to delay again. I had given notice and left my corporate job. I was all set to leave for Basic Training on September 11th, 2001.

Chapter 1

Taking an Oath

11 September 2001

I got up at 2:30 a.m. that Tuesday morning, so my recruiter and I could arrive at the St. Louis Military Entrance Processing Station, better known as MEPS, by 6:00 a.m. I've never left my family for more than a couple weeks so the goodbye was very difficult. With the anxiety of Basic Training burning inside and the ill feeling from leaving my family setting in, I felt as if I had the flu. At 28 years old, I'll swear into the United States Army and leave for Basic Training. I spent the two-and-a-half-hour ride to MEPS reaffirming that I had made the right decision. My ten-year career in computers and electronics was now in the rear view mirror, and a brand new adventure was just beginning. Upon arrival, my recruiter escorted me to the Army in-processing center with a packet of paperwork that told my life history.

Around this time, in other parts of the country, terrorists are boarding airplanes to kill thousands of innocent people—the very thing I'm swearing to defend our country from. Was my entry date a coincidence, an anomaly, or was this significant? I don't know the answer, but that thought will dance in my head forever. After sitting for nearly two hours, I asked to turn on the T.V. Every channel was displaying the terrorist attack on the twin towers in New York. The atmosphere suddenly changed, people were distracted and horrified. The staff was discussing the event. It was evident from the frantic side conversations that some had family and friends that worked in the attacked buildings. With all the confusion, my anxiety had almost completely vanished during that void of time. It was such a surreal experience, that I wasn't even processing the after effects of the attacks, and never considered that this could be a guaranteed ticket straight to combat.

After waiting nearly six hours, it was finally time to process my military contract. During the processing of my contract, the building filled with anxiety and panic, and was evacuated in fear of further attacks on government buildings throughout the country. The admissions clerk processing my contract refused to leave, saying if we left my contract incomplete it would cause major issues and delay me, so she opted to stay and finish it. She pointed out the multitude of places I had to sign, which I did hurriedly so we could exit the building. We quickly moved down the stairs. We were the last two out of the building. I am now part of the United States Army and I will undoubtedly fight this enemy that had attacked us on our own soil.

All air traffic was halted, so needless to say, I didn't leave for Basic Training that day, or the next three. I asked them to send me home and call me when they were ready for me to come back. Surprisingly, they agreed. After saying my goodbyes, I returned home only to re-live that dreaded goodbye moment a second time. It was all worth a few more days with my family.

I returned a full week after the attacks on 9/11. With no military experience, I didn't know the ramifications of the events or how I would become involved. The MEP station handed each of us a packet of information and sent the group of us out the door to a bus. We eventually made it to a city train that took us to the airport. Destined for Fort Sill Oklahoma, where the wind never stops blowing, we boarded the nearly empty airliner. We were seven of the nine passengers on the flight.

Our arrival on the ground wasn't met by yelling drill sergeants, that came after we were well within the confines of the military base. We grabbed our single duffle bag and took a bus to the barracks. Soon after our arrival, we met our first hard-core drill sergeant.

> *"Listen up! You're mine now, you understand that? I will tell you when to eat, sleep, breathe, shower, shave, piss, and shit! You do what I tell you, when I tell you and you'll make it through this hell quick and easy! You understand that? I said, DO YOU UNDERSTAND THAT!!"*

We replied, "Yes Drill Sergeant!" to everything he had to say and quickly marched off the bus. We were queued up with several other newly enlisted soldiers and brought into a holding area like cattle to the slaughter. By this time there were several drill sergeants present, all gracing us with a level of profanity that would make seasoned soldiers blush. The senior drill sergeant eventually came out to give us our in-brief. His opening statement was, "Boys we're going to war. It's our job to prepare you mentally and physically to defend this country." I felt so motivated and patriotic, but also intimidated and somewhat scared. I wasn't sure if I had the ability to complete Basic Training, much less go to war. One thing I did have was will and determination. I don't accept failure at anything, and this was no different.

Basic Training proved to be a time of physical and mental strength building. One thing was certain, when my head hit the pillow each night; I was asleep within a minute or two. They worked us like mules, day in and day out. I wrote letters home and they wrote letters back. I'm sure my absence was more difficult for my wife and two baby girls back home since this is the longest we have ever been separated. I spent a total of eight weeks in Basic Training. It would have been nine, but we lost a week as a result of the 9/11 attacks. I graduated on Wednesday, the day before Thanksgiving. My in-laws brought my wife and daughters to the graduation. My parents were unable to attend; my dad had recently undergone an operation for colon cancer, and was still recovering. It was difficult for him not to be there, but I know he and my mom were both proud parents and wanted to attend. I was afforded a single day off before being shipped out to Warrant Officer Candidate School in Fort Rucker, Alabama.

23 November 2001

Another bus to another plane and I arrived in Alabama, home of Army Aviation. On the civilian fast track to becoming a combat helicopter pilot, I left Basic Training as an E3 and became an E5 upon signing into the Warrant Officer Candidate School, better known as WOC School. After signing in, I was put into "snow bird" status. Snow birds are candidates waiting to start the next class.

As snow birds, we read regulations and studied the guidelines of the school itself. WOC School is designed to develop your ability to multitask, prioritize, test and hone your attention to detail, and develop leadership skills. Although not tailored to physical strengthening, WOC School was physically challenging for me. It was designed like all other military schools; they tear you down only to rebuild you stronger than you were. Fourteen weeks later, I graduated as a WO1 or Warrant Officer, grade 1.

March 2002

After WOC School, I travelled back to Missouri, packed the family and the house, and moved them to Alabama. We were fortunate enough to get an on-post four bedroom house beside the park, a much appreciated convenience for the kids. This was a difficult move for my wife as she had previously declined an opportunity to move a mere 90 miles because it was "too far from her family." With Basic Training and WOC School out of the way, I was finally ready to start flight school. There were so many new flight school students that I had a four-month wait just to get started. All of us waiting to get into a class were farmed out to help local contractors on the post. My job was computer operator at the air traffic control school. I worked from 6 a.m. to 3 p.m., Monday through Friday for the next four months.

June 2002

Nearly a year after signing on the dotted line, I finally started flight school. My schedule was extremely demanding. I was up by 0445 hours, and catching a bus to the flight line at 0515. We returned from the flight line around 1130 hours, ate lunch, and were attending classes from 1300 to 1700 hours daily. I went home for a quick bite of supper, and then it was off to the library until they closed at 2100 hours. I would unwind between 2100 and 2200, so I could get my mind shutdown enough to sleep. From Friday at 1700 hours, until Sunday afternoon I spent time with my family and much of that was on the sandy beaches of Panama City, Florida. The down time was a requirement I set for myself, and it proved to be a good way to offset the learning curve and this fire hose of knowledge I was drinking from.

19 March 2003

At 1000 hours, in the aviation museum, our graduation ceremony commenced. In the top ten percent of my class, I graduated from flight school and was awarded my aviator wings. Both my parents and my in-laws came to Alabama for the graduation; we made quite a celebration out of it. Strangely enough, while we were celebrating and having a great time, the initial invasion of Iraq was taking place the very same day. Is this another coincidence? I joined on 9/11, and now I'm graduating flight school on the same day we invade Iraq.

My next step was to complete my aircraft qualification course, or AQC, for my advanced combat aircraft. The OH58D Kiowa Warrior was my aircraft of choice. Better known as the 58 or Kiowa, the mission was to provide aerial reconnaissance, security, and close combat support for ground troops. The Kiowa is an armed two-pilot helicopter with no on-board crew. It's equipped with minimal armor and is typically flown without doors to improve visibility to the ground and to enable the pilots to shoot their M4 semi-automatic rifles from the cockpit, in flight. It weighs 5200 pounds at max gross weight and is very agile in flight. The Kiowa can be configured with a .50 caliber machine gun that shoots at a rate of up to 850 rounds per minute, a seven-shot rocket pod, multiple laser guided hellfire missiles, and at one time air to air stinger missiles.

There was quite a backlog of pilots awaiting the Kiowa course, so once again I was in a holding pattern for six more months. During this time, I worked directly for the Initial Entry Rotary Wing School Chief. My technical background and good work ethic found favor with the Chief so he put me to work on a few projects. I redesigned the company web page, wrote some class management software, and helped maintain database software written by another student. At one point, the Chief was getting pressure from the Department of the Army to send us to a different airframe because of a need for pilots as the War ramped up. He stayed good to his word and did not allow them to move us. Most of us were honor students and worked hard to get the aircraft we wanted.

September 2003

I spent the previous six months studying the aircraft and mission and was well prepared for the course. I won't go into the specific phases, but will say that some proved to be quite a challenge. By the time I reached AQC, we had chosen our duty assignments. I had two choices, Korea or Fort Polk, Louisiana, which was currently deployed to Baghdad, Iraq. I chose Iraq. One factor in my decision was that my unit, 4th Squadron 2nd Armored Cavalry Regiment, was scheduled to return home in March. This would give me a brief introduction to actual combat and minimize my family separation time. Yes, it was a big risk, but I didn't train for nearly 18 months so I could go sit in Korea. I wanted to fight and test my newly learned skills in real time. I needed to know my potential, to see if I had what it took. I soon found out.

November 2003

After Thanksgiving, we left for Fort Polk, Louisiana. Upon arrival we were told there was no housing available, and they put us in a two-bedroom apartment until something became available. There was nothing on the horizon so we began looking for an off-post house to rent, and again, nothing was available. After only a short period in the 2 bedroom, we resolved that we needed to buy a house and move on. We bought a small home near the post and ordered delivery of our goods. Once we signed for the house, we arranged our household goods delivery and waited.

December 2003 – February 2004

Shortly after Christmas, our household goods were delivered and we began the unpacking process. About three days after moving in, we had a flight of troops going to Baghdad and they wanted me on it. I also found out that I should have already been promoted to Chief Warrant Officer 2 back in November, but I was in transition from Alabama to Louisiana so it hadn't happened. The Personnel Office asked me to sign a new oath, and said congrats you're now a CW2— no ceremony, no people, no swearing in. Six days after our household goods were delivered, I was leaving for Baghdad. We hadn't even unpacked yet. We had the major furniture items moved in, but there were many boxes unopened. We said our goodbyes and I left knowing that it would only be two months before I returned home to finish unpacking.

Chapter 2

Welcome to Combat

14 February 2004

It's Valentine's Day, and I'm sitting quietly on my bunk in Kuwaiti waiting for the next flight to Baghdad. The tent is filled with young soldiers barely out of basic and just a few of us officers, or O types, as we're sometimes called. The room is intense. Some of the guys are quiet as mice, while others are loud and boisterous. Everyone has a different way of dealing with stress, I guess. Thoughts about what tomorrow will bring fill my head. The anticipation feels like a virus tearing through my body. I'm on this bunk fearing the unknown. What are my chances of survival? Will it be a constant battle with the enemy? Will I make one little mistake and pay with my life? All the while, I try to maintain confidence in my training and abilities. Confidence comes from experience and I had very little of that. I keep finding myself just sitting here whispering short prayers asking God to protect me throughout this journey.

Even though I know the answers, I wonder, "How did I get here?" "Why do people fight like this?" A well-worn phrase comes to mind, "The price of freedom isn't cheap." What does freedom mean to me? It really comes down to a simple concept. To have freedom is to have choices. Tomorrow we enter a combat zone. This is our opportunity to exercise the oath we took to defend and protect against all enemies foreign and domestic.

15 February 2004

We moved to Ali Al Salem Air Base to await our flight to Baghdad. A large dust storm delayed the flight for hours. The only thing to do while we waited was think about how this short trip would change our lives. During the ten hour wait, we watched the Mel Gibson movie, We Were Soldiers. Maybe it wasn't the best choice under the circumstances. Around 1630 hours, we finally departed in a C130 that was flown by the Milwaukee National Guard.

Due to the significant threat of surface-to-air attacks, the brief flight gave way to an aggressive, heart-pounding spiraling approach to the runway. After descending to a low altitude, I noticed we were dumping flares from the aircraft missile detection system. Wow! Were we being shot at? The systems aren't full proof, but very reliable, so I assume we were. After the unorthodox approach to the airport, we landed safely just after 2100 hours.

We unloaded bags and cargo and waited for buses to take us to the civilized side of Baghdad International Airport, a.k.a. BIAP. We departed around 2300 hours, and even though I could actually see our destination, it took thirty minutes to get there. The road was long and rough. We weaved through multiple checkpoints and barricades of concrete and razor wire. Our temporary home is a tent city on the airport. There are tents everywhere, all equipped with cots and air conditioning. The perimeter is lined with sand bags approximately four feet high and there are several concrete bunkers in close proximity. In the distance are several hangers damaged during the initial invasion. Concertino wire marks several boundaries. Hesco baskets, which are six-foot baskets filled with dirt and covered in razor wire, line the camp perimeter. The air is full of dust and generally keeps your skin feeling rough.

We were assigned tents on arrival. At approximately 2330 hours, when we barely had time to start moving our gear to the tents, I suddenly heard the loud whistle of a rocket scream overhead. As the rockets exploded we ran to the bunkers located just outside our tent. The violent explosions shook the ground, and were so loud it felt and sounded like they were within 50 feet of me. The closest rocket hit an air conditioner and blew it into a tent where it took a chunk out of a soldier's thigh.

Tent City on Forward Operating Base BIAP (Baghdad International Airport)

I'm sure the events of those few minutes will be seared into my mind forever. This was my first experience with a real threat. I am definitely in a war zone. It doesn't get more real than this. There were a total of three rockets launched at us; the closest one impacted just over 100 meters from me. A few minutes after the final blast a Kiowa Warrior Helicopter Scout Weapons Team flew the perimeter of our camp, and the surrounding area, trying to locate the attacker.

After the blasts, I sat on my bunk in shock. I couldn't help but think of my family and the real possibility of death. I may not go home. I told a fellow pilot going to my Air Cavalry unit, I would sleep in my flack vest and Kevlar if I had to, but I'll be damned if I'm going to die here!

Bunker outside our tent

16 February 2004

I didn't sleep at all last night. The fear of another attack had me too on edge. Between thinking about what might happen during my time here, and praying for my safe return home, my mind just wouldn't stop. And the tragic and terrifying events keep coming. Our leadership team conducted an emergency accountability formation because they found someone murdered and wired with explosives on our base. It turned out to be an Iraqi worker. I'm glad to hear it wasn't a soldier but it reminds me that we have enemies in our camp. To prepare us for inevitable events, we were required to attend classes explaining in explicit detail everything this combat zone is sure to bring to our lives.

17 February 2004

Our camp has already fallen into a routine—we are up at 0545 and off to chow, get all the bad news from the night before, and then move on to the rest of our day. One-hundred-twenty new soldiers arrived at BIAP last night, swelling the size of our camp. The high number of troops makes us a good target for attacks.

Today at 1600 hours, we sat in a medical class learning how to treat some of the many wounds we are sure to encounter, as well as learning how to react in traumatic situations. Almost as if on cue, the building was rocked by a very large explosion. The windows shook and the ground rumbled as we quickly moved away from the windows. This attack made the first one seem insignificant. My heart was pounding. When it was over, we learned that several mortars had struck somewhere near the Air Force' section of the airfield. This attack wasn't even close to us, but it was still the loudest explosion I'd ever heard. The intelligence group predicts we will experience another large mortar attack in the near future. The word is that it will be approximately 20 rounds. The norm seems to be 3 to 6 rounds. I pray God will watch over us, and protect my tent and this camp.

18 February 2004

While we were out completing our training today, the 1st Armored Division fortified the sandbags and bunkers around the tents in expectation of a larger attack. It's inevitable that we will be attacked. I've accepted that. It feels like a game of chance. It's pointless to constantly worry about it. I refuse to live in fear. I'll do the job I came to do and won't be scared. If I die, I die, but I'll do it with honor and without fear.

I signed up to use the satellite phone, more commonly known as the "sat phone," at 1810. I've thought about my family a lot this past week. I've also thought about how odd it is to be in a war where my mission is to help the Iraqi people, even though some of them would kill me without a single regret, all the while believing they will be rewarded in heaven for doing it. It's unbelievable. So why am I here helping people that are trying to kill me? It doesn't seem right, but I suppose they have been oppressed for so long that they're brain washed. Whatever the case, I've learned that the first thing you should know here is that you can't trust anyone except your brothers in arms. The enemy will kill you so fast and not think twice. I'm always watching my back, always feeling like someone is after me. If you don't intimidate them right away, they will target you, and wait for an opportunity to kill you. Don't trust anyone.

Tomorrow we'll be picked up by our receiving unit and convoy to our respective Forward Operating Bases, or FOBs. Convoys are very dangerous because they're easily ambushed. This will be another new experience. From the talk around here, my unit, the 2nd Armored Cavalry Regiment, doesn't mess around. If someone acts threatening, they eliminate the threat. While I look forward to getting to my unit and doing some flying, I dread the risk of the convoy. I've heard horror stories of ambushed convoys and improvised explosive devices (IEDs) being used to stop convoys, damage equipment, and kill people.

> *I'd like to say how much I love my family and miss them. Lord please protect me while I'm here in this hellish place. I pray your angels of protection around me and my quarters. I also pray for protection for my unit and their convoys. Lord bless my family with my life and my safe return, amen.*

19 February 2004

Traveling on the Iraqi interstate was very different than zipping down an interstate back home. Just before we rolled out of BIAP one of the sergeants dumped about 30 M4 mags in the back of the truck along with a few body armor plates. We loaded and readied our weapons. As we left, my heart was beating like mad. My adrenaline was pumping like never before. There were several gun trucks traveling with our convoy. The serious and focused look on each soldiers face showed me they were seasoned and ready for the mission. We were sure to overcome anything we encountered. Driving on Iraqi highways was more like bumper cars on the interstate. The lanes aren't marked and people tend to drive only one or two feet apart. If they bump another vehicle, oh well. They certainly didn't mind driving close to our convoy. I felt vulnerable because it was so different. On the last convoy I rode in the right seat in the back of a Humvee. I wasn't assigned an M4 rifle so I un-holstered my M9 pistol and carried it loaded and ready until we were safely inside Camp Muleskinner.

Along our route we encountered a local man who had killed a lamb or goat by decapitating it. He held its head up chanting something and shaking the head at us while blood ran from it. Iraq is a filthy and unsanitary place. Trash litters the streets. Sewer systems only exist in some parts of the inner city. In other areas urine runs out of homes in small trenches to the alleyways. There are very few trees and little grass. It's not a pleasant sight. I see farmers up and down the sides of the road farming by hand. I'm amazed at the difference between their lifestyle and ours. Americans have no idea how good we have it.

*Just before departure – Convoying from BIAP
to Regimental Headquarters then on to Camp Muleskinner*

Chapter 3

Camp Muleskinner

19 February 2004 continued

After surviving our first two convoys, we arrived at our Forward Operating Base, Camp Muleskinner, at approximately 1300 hours. At the entrance checkpoint, we were required to exit the vehicles and safe our weapons. This was a standard practice upon arrival at any base and prior to entering many facilities on the base. My platoon leader escorted us to our respective troop areas.

Our assigned living area was an open forum in a large one room building, also known as "the hooch." We had approximately 30 people living together in that area, each with a small area to call their own. Several people built walls and bunk beds from plywood. Some went the extra mile and fabricated desks and shelving units. My "room" was an 8' x 6' area with a cot and no walls. I quickly sprung into action, stringing up support runners made of 550 cord to support hanging blankets and ponchos. An ounce of privacy goes a long way. Our building had been recently renovated to include a shower room/ bathroom with plumbing. Before this upgrade to the building, there was no running water or plumbing, so feces had to be burnt and showers were gravity-fed sprinkler heads with ponchos strung up for limited privacy. I'm thankful I didn't have to live in those conditions.

8'x6' Living Area

Once Saddam's War College, Camp Muleskinner was an interesting place. It was strange to think that not long ago; Saddam trained his officers, unsuccessfully, to defeat us. Now I'm sitting in his nest, we were the wolves in his hen house. In the spare minutes I had, trekked around to different buildings to absorb the history and imagine the activities that had taken place. Many of the buildings were bombed out and not safe to be in. I stood in the back of the movie theater and looked at the collapsed ceiling and pictured the room crowded with Saddam's men, sitting in these very seats, being briefed on the engagement that would eventually lead to their defeat.

There was an Olympic-sized swimming pool and several training buildings. Many of the classroom buildings had been converted to shops where we could get a few convenience items such as candy, soda, chips, and even some locally-made jewelry. I'm not sure what kind of candy bars they were as the labels were written in Farsi. I sampled a few, but found only one that was favorable to eat. They were selling homemade foods until the soldiers became ill from eating them. We were quickly banned from eating anything that wasn't a packaged product. They had some pool tables in the place, which proved to be a much needed escape for us during our off time. I snapped several pictures of the campus and destruction for my own record.

Our assigned living quarters on Camp Muleskinner

Destroyed Movie Theater

20 February 2004

Mike and I went to the flight doctor and he signed us off for flight duty. Upon arrival to any new unit, you're required to see the flight doc and be evaluated and "signed off" for flight duties. I am slowly getting my mandatory in-processing duties completed. Military paperwork never ends. Even in combat there's a pile of required paperwork. At 1300 hours we went to the airfield with an instructor pilot and went through a pre-flight inspection on an aircraft. I think I'll really like my instructor pilot or IP as they are more commonly called.

I took my third anthrax vaccine today, and that wasn't fun. It feels as though you took a knuckle punch from boxer. The soreness drags on for several days. It doesn't matter if you don't want it, if the Army says you're taking it, then you do it with a smile and move out smartly. Remain mission focused; you're a machine and nothing can stop you. Hoooah!

I'll be studying more intensely since we're on the rigorous two-week progression track. Readiness Level progressions are the phases of training a pilot receives upon arrival at his unit. This progression would normally take three to six months stateside, instead of two short weeks here. Ideally, each day we would fly about 2 hours. Another progressing pilot flew 10 hours today, this was an exception, but we'll be training hard and be combat ready as soon as possible.

I called home today; it's always recharging to talk to my family. Hearing voices from home keeps me motivated and gives me a short lived escape from this hellish reality. My days seem short because time flies when I'm busy. It's important to become the best pilot I can be, and no matter how much I work at it, I still know I can improve. There's so much to do in very little time.

On the flight line (indoor olympic sized pool pictured in the background)

I'm integrating well, but still have some uniform patches to get sewn on, in order to look like everyone else. We have a "new comers brief" tomorrow at 1000 hours in Flight Operations. How many briefings can one get? It seems like we never stop getting briefed! I just want to start flying and doing my job. Well its lights out. I am so thankful I'm no longer at BIAP. The attacks here at Muleskinner have been much less frequent than BIAP, thank the Lord.

21 February 2004

I got up around 0700 hours and went for chow. Our morning briefing lasted approximately 1hour and 15 minutes. We then went to draw our basic load of live ammo to complement the mere 5 rounds we had been carrying, followed by a short link-up with our IPs. After talking with the IPs for a while, we went back to our quarters, or "the hooch," and picked up our Aviation Desert Combat Uniforms, a.k.a. ADCUs. With our new uniforms in hand, we went to a sew shop on the FOB to get our uniforms standardized. Each FOB allowed vendors to setup a sewing shop to accommodate the needs of the soldiers. After the uniform drop we ate the noon meal and learned that the squadron commander wanted to go on a run with each of us next week. Wow! I'll have to do some training for that one. I hear he normally runs about 5 miles or so. That's about an hour's worth of chit-chat.

The "Green Platoon" as we were called, met with the IPs at 1400 hours and discussed the demanding progression schedule. This would prove to be a tough two weeks, I was sure of it. The birds we would be flying were older models than the birds we had trained in at the school house back in the States. We had trained in the OH58D-R and would be flying an OH58D-I. While not greatly different, there were a few differences and would require training to bring us up to speed. We went to the flight line, where the aircraft were parked and did some routine tasks to familiarize ourselves with the older model. Then it was off to the chow tent for dinner, followed by some bunk time. The much needed rest felt great. With such a mental and physical work load, I could fall asleep as soon as my head hit the pillow.

22 February 2004

The next morning Mike and I went to a small arms range on the FOB to demonstrate proficiency with our M9 pistol. Didn't we just do this back in the States? Ya gotta love the Army, we had to jump through hoops even while deployed. The run with the squadron commander was quickly sneaking up on me so I went for a long slow rehearsal run. I sure didn't want to be the guy who couldn't keep up with the old man. I washed some clothes in the make shift laundry facility then went to "ha-beeb's," internet café, to send a few emails. Yes, they did have some internet service, but it was intermittent at best.

On the way back, I heard two loud explosions. They weren't close, but they were not Explosive Ordinance Detachment (EOD), which were controlled detonations at the top of the hour just outside the FOB. Muleskinner was being attacked every two to three days, but I haven't had any real close calls since my arrival and they're much less frequent than during my time at BIAP.

We'll go eat chow at 1200 or so, then it's back to the flight line at 1400 for training. We went to the SRP, or Soldier Readiness Process, for redeployment back to the good ole U.S.A. This was a great feeling! I had only arrived and we were already discussing going home. The briefing was about an hour, and then it was off to dinner. After dinner we went directly to the TOC, or Tactical Operations Center, for more classes. Our instructor pilot went over a few topics with us. It was mainly review, but I did learn a few things. While we were in the class, we heard an explosion in the distance. Again, it wasn't too close, but this one was close enough to knock a mirror off the wall. I was slowly growing more accustomed to the attacks.

We wrapped up at about 2130 hours and called it a night. Tomorrow we'll go for a Performance Flight Evaluation, so tonight I'll be looking over all my basics one more time. Hopefully, I'll perform well under pressure tomorrow.

23 February 2004

We worked on our aircraft operator manual exam, or "-10 test" as it's known, followed by some base flight tasks in preparation for flying that afternoon. My nerves were on edge as I prepared for my maiden combat flight. As if being in hostile territory wasn't bad enough, I'm also being evaluated in action. The Army school house instructors always said "Don't worry; rely on your training and instincts." My instincts were telling me it was dangerous, and to seek shelter, but someone has to do it. I volunteered, and here I was in the thick of it, so let's get this party started. Hoooah!

My first flight went well. Even though I didn't feel as threatened as I thought I would, it was obvious, we were in a combat zone. Many buildings were riddled with bullet holes and some were in rubble from the destruction of the initial invasion. There were disabled Iraqi military vehicles here and there. From the air, the city was a color picture out of my grade school Bible. Rasheed Airfield was right next to our FOB and would be our playground for the next couple days.

The airfield wasn't within our FOB boundaries and was peppered with local Iraqi citizens. Some were bringing their cars to the stale ground water and using it to wash their cars. Rain was rare in Baghdad and the Iraq water-shed was non-existent so water flooded the airfield after a rain. It sounds strange, but they have little water so this is a standard practice. They seemed to enjoy seeing us fly and cheered us on as we completed each maneuver.

The first flight mainly consisted of emergency procedures training and was conducted close to our FOB in a relatively secure area. We returned without issue, flight one was complete. Tomorrow, we're flying nights so we'll see how that goes.

Tonight some Blackhawks took fire from a surface-to-air missile. No one was hit or injured, but the resistance is definitely trying to ambush us in any way they can. This is obviously disturbing news and adds more tension and stress to the anticipation of my normal routine. It feels like I'll be rolling the dice everyday with my life at stake.

Daily I become more conscious of my surroundings and catch myself evaluating everyone I see. Everyday our intelligence group is briefing us on incidents involving corrupt locals; it's making me paranoid about every non-American on the FOB.

I'm excited to call home tonight. It's always nice to hear my wife's voice.

Rasheed airfield (notice the submerged helipad behind the bus)

24 February 2004

Tonight will be our first night flight in the combat zone. I was a bit nervous and needed some time to reacquaint myself with the techniques of night flight. Even with night vision goggles, night flight is a challenge and very dangerous.

I guess I did okay for not flying with goggles for so long. Night vision goggles allow us to see in the dark but there's a lot going on in our cockpit. I have 4 radios blaring in my ear, I'm trying to maintain position in the formation flight, dodge obstacles such as towers and power lines, all while trying to complete the mission of protection for a ground team or reconnaissance to find the enemy. It's very intimidating for new pilot.

We flew about two hours and covered all the required tasks. We were originally slated to fly an actual combat mission, but Lee, my IP, and I agreed that it wasn't the most brilliant plan so we scratched that and flew later in the evening. That mission would be executed by a more experienced crew. Lee is a great instructor and demands performance.

I've done well so far and am happy with my performance. I see things I can improve on, and I will. I will be one of the best.

25 February 2004

I was up late, but still managed to rise before 0800 hours. I have several things to study for my flight tonight. I'm going to the finance office to grab some cash. I want to pick up a tea set from a local vendor for my girls. They'll enjoy getting it.

It's good that we switched up missions last night because they had some action and I don't know how much of an asset I would have been without being mission ready. The last thing I want to do is cause casualties. Tonight will be my first real combat mission over Baghdad. Am I ready? What should I expect? How will things go? How will I perform? These are all questions that filled my head.

I met my IP at the flight line; we did our pre-flight on the aircraft, received our required intel download from the S2 Officer, then headed to the aircraft. We strapped in and with all pre-start checks completed, he hit the start switch. The starter whined, followed by the roar of that Rolls Royce jet engine. The smell of jet exhaust burned my nostrils as we readied all systems for flight. In a mere 3 minutes we were ready for lift off. My heart was pounding as we cleared the perimeter wall and armed the weapon systems. Most of this flight would be conducted at altitudes below two hundred feet off the ground. Weaving, climbing, and diving, we made our way through the city to our mission area. We conducted a route reconnaissance and patrolled the routes for suspicious activity. It was also an opportunity to become familiarized with the local area and learn the aerial hazards. No matter how many times you see the area in daylight, it always seems confusing at night on the first few missions. This was a great training mission since we met no enemy resistance.

With our mission complete, we returned to the FOB. After engine shutdown, I climbed out and felt pretty good. At first I had some anxiety, but I was able to adjust within thirty minutes or so. It felt good to finally be an asset and add value to the team. The mission took approximately three hours. Many of the aerial hazards are unmarked. Obstacles such as towers and power lines aren't lit like they are back home, making it very dangerous at night.

We had no incidents and I quickly became acquainted with the local area and the type missions I would perform on a routine basis. I was able to do my job as trained. We went into the flight operations office and debriefed the intelligence officer on all that occurred during the mission and any suspicious activities or oddities before calling it a night. My IP gave me a mission to plan for tomorrow night so I guess I'll work on that tomorrow morning, it should be no trouble.

I think of my family during my down time. I really miss them and hope to see them soon. I love you girls.

Typical Iraqi homes – notice the sewage trenches that extend from the perimeter walls

*Most of these trenches are between the houses
and away from the main streets.*

Chapter 4

Progression Training

27 February 2004

We're prepping to go to the gunnery range and will receive our operations briefing today. I can't believe we're already shooting gunnery. It'll be great to be shooting again. The smell of the gun powder, the vibration of the 50 Cal two feet out the door, and the whoosh of the rockets leaving the tube—let's get some! I had a great time shooting back at the training range in Alabama, so the real thing should be even better. With all the recent attacks, I'm ready for some payback. I hope my aim is as good as my confidence.

I flew until about 0000 hours last night, so I took my time getting up this morning. I have to be back at the airfield at 1400 hours for more fun. I missed writing anything yesterday, but managed to send a few emails and get some cash for gifts. I'm buying my girls a tea set. They love tea sets and I love surprising them with special gifts. Buying them gifts seems to be an emotional outlet while we're separated. I'd like to share more with Roberta about how I'm feeling, and what's going on here, but I can't really talk about it for security reasons. I love her dearly and hope to see her soon.

Staying busy is the best distracter and keeps me from dwelling on missing my family. I wish I could hug my girls and say "I love you." The military will definitely test the strength of your family and the bond with your spouse. Many of the soldiers have been here for a year, some marriages survive, and others do not.

The soldiers are burnt out and ready for some well deserved down time. We work 7 days a week here; it's not like back home with weekends off. My hat's off to the group that's been here so long. Outlaw, my assigned Troop, will leave Iraq earlier than most because they deployed a month earlier than the majority of the squadron. With my late arrival to the squadron, because of training, it was decided that I would stay in Iraq and fly missions with Palehorse Troop. I have no objections to this and actually prefer it to gain experience and build flight time. I'll transition home with the main body, which is scheduled to occur in a month or so. I should be home within 2 months from now. *"God keep me safe in this journey that I may safely return to my family."*

28 February 2004

Last night I heard an improvised explosive device (I.E.D.) explode around 2130 hours, and from the news around camp it was a planned attack on the Regimental Commander's convoy. Two soldiers were wounded, but fortunately, none were killed. Every mortar or rocket attack I survive makes me feel like I'm playing roulette with the highest stakes, my life. The insurgency is ruthless. It's difficult to justify helping the Iraqis because some are just murderers. I have to stay focused on the ideal that I'm protecting my fellow soldiers and making our country a safer place. Most Iraqis are grateful for our presence. Yesterday while flying missions, all but one person waved at us. I got a "thumbs down" from one man so I gave him a friendly dose of dirt in the face with my rotor wash. Ha! Most are grateful for our efforts. I've not seen any local news from the States so I don't know what goes on outside my small circle.

I took some personal time to escape my reality by writing a few emails home today. It feels good to have the ability to communicate with my family. I bought a blanket for my wife today, and tomorrow I'll get the tea set for my baby girls as well as some Iraqi money for souvenirs. The Iraqi currency was replaced after Saddam was removed from power. Some of the guys bought a bunch of Dinar in hopes it would gain value some day.

The exchange rate was pretty high. I can't remember exactly, but seems it was somewhere near one thousand to one. I also picked up a few DVD movies. They're all copies coming out of Turkey. Some are good quality, while others are cheap, but overall not bad. I need to mail that stuff home to the girls but right now it's time to go burn some jet fuel!

29 February 2004

I picked up the tea set for the girls and some Iraqi money for souvenirs. We took the Helicopter Gunnery Skills Test today, which is the written exam required prior to shooting gunnery. I passed, so it looks like we'll go lay waste to some targets soon. Whew hoo! We'll be shooting the real deal this time, high explosive rockets, no more trainers.

Today brought some excitement. EOD destroyed some explosives recovered during day to day operations. By now I'm well used to them blowing charges daily around 1600 hours, but no one was prepared for this one. When it detonated, I thought we were being attacked! I was in a room that had a door open on the same side of the building as the explosion, and I was in another doorway on the opposite side of the room. When this thing blew, it pushed me out the door and rattled the building so hard that the metal vent screwed to the wall was pushed out and hit one of our Captains in the shoulder. The power of this explosion rocked the city. All of the windows in the command center were cracked or broken out. The blast broke windows for several blocks. The new gym was a disaster. All the windows broke and the drop ceiling came down. In the aftermath of the very powerful, but controlled explosion, I couldn't help but think that I sure wouldn't have wanted to be on the receiving end of our initial invasion into Iraq. We use ordnance much more powerful than this.

EOD detonation from the above description.

Detonation was staged inside a concrete bunker for additional protection.

While out flying a night recon mission, we got the call to return to base, so we high-tailed it back. While we were away, the base camp had sustained another mortar attack. Thank God they didn't hit inside the perimeter. Once back at camp, we conducted a perimeter sweep and provided security against follow-up attacks. My blood was pumping and my heart was pounding in my chest. Thoughts rushed through my head. Would they try shooting us down? Would this be my first trigger pull on the enemy? I'm now hunting people, something very new to me. We did find something that gave us insight on an enemy location. Even through the danger and threat, I was actually doing what I was trained to do, and it felt awesome!

I was flying left seat, which inherently has your focus inside the cockpit with your head down using the Mast Mounted Sight (MMS). While using the sight and circling multiple times, I did get a little nauseous. I hope that goes away with time. I ended up with about 4 hours of flight time and experience. I fly a lot and that's what it's all about; getting up in the air and doing my job. As long as I come back safely every day, I say it's a successful day.

1 March 2004

I slept until 0930 because we've been up late nearly every night. I'm slowly adjusting to the night mission. It's still a very eerie feeling to fly above hostile ground in the black of night. There are very few safe places to land in our operations area. This feeling is also referred to as "the pucker factor." The more intense the situation, the more your butt puckers with nervous intensity.

I still have a week or so of progression training remaining. We'll go in at 1300 today in preparation for an early rise for tomorrow's gunnery. We'll be riding out on Blackhawks and will spend the day at the range. I finished the Readiness Level 3 (RL3) tasks today and am now RL2. After gunnery I'll work toward completing RL2 and being fully mission capable.

I practiced pinnacle landings today on a rooftop that was barely wide enough for the skids to fit on. With pinnacle landings, you land on a small point, typically on a ridge or mountain top, but Baghdad consists mainly of flat terrain so we simulated it using the small rooftop of an abandon building. It gets real interesting when you can only see your touchdown surface through the bottom of the aircraft. Next I'll have to do the same using night vision goggles, which will be challenging.

It feels great to be finished with the first phase of the progression training. I'll celebrate by relaxing for a bit and watching a movie. I watched "Behind enemy Lines." It seemed appropriate for my given scenario. Sometimes I think of what I am doing, and it's odd that there are people all around me who want to kill me merely for what I represent. It's difficult for me to imagine a lifestyle as violent as theirs. The Iraqis are very poor and oppressed. It's all they know. It's their way of life.

2 March 2004

I was able to call home again and talk to my wife—what a wonderful treat. It's recharging to hear her voice. Our Squadron's gunnery starts today, but my shooting festivities will have to wait until tomorrow. Mike and I were tasked to work at the range and help load ammo as the birds rotate through the fuel point. It should be an exciting new experience.

3 March 2004

We started gunnery and shot all our daytime requirements. I am getting very familiar with the aircraft, and flights are becoming more fun. I had to hitch a ride back to the FOB on a hawk because our aircraft was having issues with high engine oil temperature in the desert heat. After returning, we picked up another aircraft. I flew us back out to the range and it was exhilarating, fifty feet off the ground at 110 mph. We fired more during the daylight hours, but the night fire was cancelled due to smoke from a local factory blowing across the range and causing visibility issues. The smoke was accompanied by the nasty smell of sewage. There's always feces, smoke, or some other unidentifiable smell lingering over the city. With no EPA, I often wonder how much of this is harmful to my health and if it will have lasting effects.

4 March 2004

I mailed the gifts to the girls today and it only cost $22, not too bad. I was prepared to pay more. I was flown by Blackhawk out to the range at 1730 and completed my night gunnery training. The Army never disappoints in their venture to hurry up and wait. I sat from 1730 until 2300 before I was able to shoot. There were several pilots completing their annual range requirements, even while deployed to combat. It seemed strange to me, but I just do what I'm told. By the time we completed and returned to the FOB, it was 0230. It's about 0330 now so I'll undoubtedly sleep in.

Tea set purchased in Baghdad from a local merchant on our FOB.

5 March 2004

I woke up around 1030 today and was still tired from last nights' festivities. We had a show time of 1500, so I had a few hours before I had to go to work. We ate chow and went to the PX before going in. When we got there, our instructors said we were not flying today, so take the day off. Wow, that was a bit of relief, now I can catch up on some much needed rest. We called home and found some internet access to send a few emails. The internet service was via satellite and very unreliable. It took some work to find a PC that actually worked. I watched a couple movies and thoroughly enjoyed the down time.

Tomorrow we fly, but Sunday is a safety stand down day, so we are off yet another day. Monday and Tuesday we'll hit it hard, and Wednesday we're scheduled to take a check ride to move from RL2 to RL1 status. Wow! RL1 already. They want us mission ready as quickly as possible. I'll catch some shut eye now and write more tomorrow. I've been thinking a lot about the family and really miss them. I love my girls and miss them dearly.

6 March 2004

I woke up around 0730, linked up with Mike, and then went to chow. Afterward we were going to study, but when I returned, I discovered that my Platoon Leader had scheduled me to run with the Squadron Commander in two days. Knowing that our commander has a passion for running, I thought I should go for a run and ensure I was up for the one-hour running interview he graciously extends to all new officers. I ran 45 minutes while listening to some high-paced workout music my wife used in her rigorous aerobic instruction course. This triggered thoughts of her. I really miss her a lot and it causes me to think of all the little things I took for granted, such as just spending more time with her and the kids.

Strangely enough, I'll be home pretty soon. I'll be taking my check ride in a few days, just in time to complete a few mission flights before our scheduled departure from the combat zone for Kuwait. Two weeks after that, we should be safely back home. Time seems to go by fast when I look at the big picture, but seems slow when I think of my family, missing Valentine's Day with my wife, or just a weekend trip to the park with my babies.

Tomorrow we are off work and I'm looking forward to a day of no obligations. Mike and I will probably still spend time studying for our check ride. The internet is down, so I've been checking it often and hoping someone fixes it soon. Bringing a PC has been a real morale booster, not only for email but to watch movies and just look at family photos.

I flew my mission tonight. It was 3.3 hours and brought no surprises. I did well and remembered more than I expected. Mike and I were assigned a lengthy mission to plan for Monday, so I'll work on that tomorrow.

I called home tonight and as always, it was good to hear the girls' voices again. My daughters were getting ready for a birthday party, which meant to me their minds were occupied and distracted from thoughts of me being away. It's the little things that keep me smiling.

7 March 2004

I slept in until about 1000 hours, and then watched part of a Jackie Chan movie before washing a bag of clothes and getting ready to move to another building on the FOB. We planned our mission after lunch which took us about three hours. Afterward, Mike and I studied aircraft emergency procedures, all the while hoping we would never have to execute one. The thought of being shot down was uncomfortable enough without the stressors of something failing on the aircraft and having to execute emergency procedures.

The weather is slightly cooler today. It's windy with an overcast sky, which is something I've not seen since I've been here. Iraq has that Arizona type of weather—hot and dry with an occasional rain storm. Someone mentioned they felt a few sprinkles earlier, but I'm betting it's just wishful thinking. It's so dry here, when the wind blows, the dust kicks up like fine powder and lingers as smoke would. It's a very dusty and dirty environment. Dust covers everything; that's the main reason we have to clean everything before bringing it back into the States. Small parasites attach to personal items and equipment to hitch a ride. This is one of the risks in moving people and equipment across countries.

I'm scheduled to run with the Squadron Commander tomorrow. It may be interesting since I'm not much of a runner and he loves to run. He seems like a good person on the surface. I haven't really talked with him yet but as the old saying goes, only time will tell. I've been contemplating our next mission and hope it goes smooth.

Mike and I plan to watch a movie tonight and unwind a little. The internet is finally working again. My wife was online so we chatted for a long while and I also talked to my sister which was a nice surprise. Talking to my wife and listening to her soft voice makes me miss her even more, but I can't stand not hearing it. It's a catch twenty-two.

8 March 2004

I got up early for my run with the Squadron Commander, shaved, brushed my teeth, and even stretched a little. We have an early mission time so it will be a mad dash to get ready afterward.

I found our Commander to be a very down-to-earth guy that loved the Army and our country. We discussed some of the pros and cons of our situation and our core values as individuals. I enjoyed the run and felt like he not only wanted to get to know me, but also test my physical ability and make sure he was getting a quality officer and pilot. He wasn't disappointed. I didn't sugar coat my opinions or thoughts and wasn't nervous about being around him. He appreciated my candid approach and we were off to a great start.

I flew a day mission today. During the mission I took several pictures of the local neighborhoods and people. It's neat seeing how the Iraqis live and how their world is so different than ours. They lack very basic things such as sewer systems and clean water. All the houses are built from mud or poor quality concrete. Houses outside of the city don't usually have doors. Running water is limited; most houses have water tanks on top of their roofs that are filled by a water vendor delivering it via truck. With the extreme desert heat, by mid-afternoon the water is scalding hot as it comes from the tank.

The flights have been a great learning experience. Even from the air we are able to tell what groups of people appreciate us being there. The locals that love us dance and wave as we fly over, others pretended we don't exist or shake their fists in anger. Our mission ended without incident.

I called home and talked to my wife today. She was doing well. The car radio died so she had that fixed today. She's feeling the pressures of taking care of everything I would normally deal with. It's definitely making her stronger. I miss her badly, and wish I was there to take care of her.

9 March 2004

I was up by 0700 hours and went to chow. I began packing up my stuff to move to another building. I washed clothes in preparation for packing the shipping container to send home. I did a little studying and bought some more movies. I've enjoyed my country music and the Jeff Foxworthy CD has been some much needed comic relief.

Mike and I flew 5 hours tonight and passed our RL1 check ride. I performed really well and was able to answer all the questions asked of me. I'm now fully mission ready. This is really where the learning starts. I now have a basic understanding of tactical combat flight and how to employ the aircraft in given situations, but the real learning comes from other pilots that have experience. I'm excited to jump in the cockpit with these veterans and soak up some real knowledge. I have nearly 40 flight hours of combat time to date, all of which has been training with low stress missions.

I was able to call home and talk to my parents today. Mom limited her questions to non-combat related topics such as how the girls are adjusting back home and what our living conditions were like. Dad wanted the meat of the heat and asked if I had been shot at. I told him not as far as I know, at least in the air. If so they missed. I told him we had sustained several attacks by mortars or rockets targeting the FOB. He asked me about the combat zone in general and I told him of my experience at BIAP. That was a scary few days. I felt the concern in his voice when he went silent. Maybe I shouldn't have told him. We ended on a good note and I was sure to say I love you to both of them because I never know what tomorrow may bring.

Chapter 5

Learning the AO

Reflecting on a few missions, I now realize that the Iraqis are shocked and in fear because of an overthrown government. Our soldiers occupy their streets and they are uncertain of their country's future. While on a mission yesterday, we flew over a guy who was walking with his rifle. As we circled his position, he made a poor attempt to hide the rifle behind his leg. The look on his face was like that of a child whose been caught with his hand in the candy jar - busted! He displayed no mal intent so we left him alone and reported it during our post flight debrief with the intelligence officer. At this point our ROE, or Rules of Engagement, allowed for public ownership and the use of weapons. This would later become limited to home use only.

Mike and I had previously visited the local market to purchase a couple of the finest Cuban cigars they had to offer to properly celebrate passing our RL1 check rides. Mike is flying so the celebratory festivities will have to wait.

11 March 2004

The next two days are non-flying days for me, so I'll use them to relax somewhat and move into my new living area. I've been living in an open bay with my troop but with our redeployment to the States only a few weeks away, Operation Break Down is well underway. I'm moving to a room in the firehouse with my Platoon Leader who also joined the unit after initial deployment. He will stay behind with the rest of us. This new room affords much more privacy compared to my current situation with walls made from 550 cord and poncho liners.

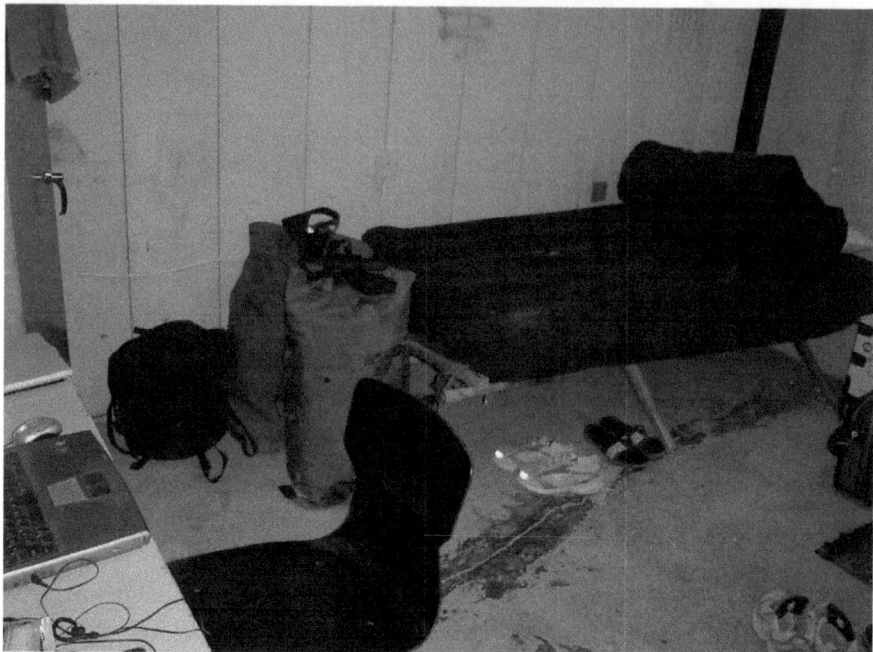

New living area inside the firehouse

Outside view of the firehouse

M1 tanks staged just outside our living quarters

My platoon leader had requested that I attend an aircraft maintenance leadership meeting and the Troop Update Brief for Palehorse Troop on his behalf. The maintenance meeting consisted of aircraft maintenance information specific to the health of our aircraft as well as the maintenance plan during the return trip to Kuwait and eventually the U.S. The Troop Update Brief, or TUB, is a weekly meeting held by the Troop Commander to disseminate information to all soldiers within the troop and consisted of requirements for each soldier to complete prior to their return home. I took copious notes for my Platoon Leader and learned some of the red tape in the process.

Until now, I've flown every mission with an instructor pilot but I was just briefed that I'll be flying my first mission tomorrow night without an instructor. All military mission flights are conducted with two pilots, one designated as pilot in command or P.I.C. and one designated as pilot or P.I. Even stateside, it takes several flight hours and much experience to be recommended to serve as pilot in command so I had no expectations of making P.I.C. anytime soon.

I had no issues with this at all as I'm in the infant stages of my new career. I was very humbled by the experience that surrounds me and had great respect for the crews that conducted the initial invasion. Flying at low levels over the city at night wearing NVGs can be extremely dangerous. I've had the best training and am with highly experienced aviators; for that I am thankful.

We're conducting a route reconnaissance in the black of night at an altitude between fifty and two hundred feet above the ground. Flying just above the power lines in the city is enough to max out your pucker factor.

I was able to call to home today. I try to keep my conversations focused on home and pleasant things in hopes of distracting them from thoughts of the very real danger I was in. Dad was preparing for a fishing trip with a couple of our friends. As the conversation labored on, I was jealous of the things I had given up to be here.

12 March 2004

I was lazy on my cot until about 1000 hours. Working nights has made sleeping in part of my normal routine. I'm usually rudely awakened by the sound of loud speakers of Mosques throughout the city playing the Muslim call to prayer. Waking up to this dreadful noise was an immediate and unwelcomed reminder of where I was.

Our FOB hosted a bazaar today and I bought both of my girls a Sterling Silver necklace. Both necklaces have a heart pendant that is hinged at the bottom tip of the heart and it opens to a display which says "I Love You." This is very much a symbol of a deep emotion; it's painful not seeing them. Buying gifts for them seems to ease the pain and provide some therapy for me. I was able to chat with my wife for a short time on the computer. I'm impressed with how well she's doing, considering the fact that I had to leave only a few short weeks after moving to Louisiana. Her parents live ten hours from us but would drop everything if she needed something.

I flew a routine four and a half hours today. We had nothing too exciting to report. It was a great opportunity to practice skills and become more proficient with my routine. I'll be on QRF duty tomorrow night and the next one as well. QRF or Quick Reaction

Force is a duty for quick response to incidents that occur in our operations area. In this case, it's Baghdad. Our mission is to provide close combat air support for ground troops under enemy fire.

I went to midnight chow and what a surprise - they had seafood! The aroma of shrimp and lobster filled the tent. With my eyes closed, I imagined I was back at home, inside a Red Lobster, and briefly escaped my reality. Wow, what a treat it was! After being spoiled with seafood from Halliburton, it's time for some shut eye.

13 March 2004

I was on QRF tonight. Luckily, the night proved to be incident free so we burned some time watching a movie and relaxing. I use the term relaxing loosely because at this point in the war, the air QRF rarely get launched but the nervous burning is still in my stomach. The anxiety of flying into an unknown situation is an ill feeling.

In my down time, I've been developing an electronic logbook to track flight hours and requirements for all our pilots. It's coming along well. This project helps to keep my mental gears turning when I'm not flying. I was deeply consumed by the project and didn't get to bed until 0230 hours.

14 March 2004

I woke up around 0930 hours or so. I'm slowly learning the processes and procedures for the daily grind and becoming more comfortable with life in a war zone. The mortar and rocket attacks have become less frequent. We're currently being attacked about twice a week, much better than the daily attacks at BIAP.

I have QRF duty again tonight and I've been given the task of flying air support for a departing convoy at around 0230 hours. We have a large group departing for Kuwait early in the morning. As I'm told, it's very rare to move a convoy this size without meeting some enemy contact. I tried napping this afternoon but was unable to sleep for my nervousness about tonight's mission. The learning curve is tremendous.

15 March 2004

Last night's mission went off without a hitch. I was praying for divine protection and had the family praying as well. We escorted over 100 vehicles out of Baghdad without a single bullet fired, which I believe to be a miracle. We took the convoy about fifty miles south to Babylon and shut down for fuel. We saw one of Saddam's palaces there. It lacked the luster I would expect from a palace but did have the contour of Middle Eastern architecture. I wish I had brought my camera and captured some of this history but the mission was weighing heavily on my mind.

It's interesting to have experiences in an area that has such a great historical meaning, even if it is under wartime conditions. This isn't the first time Babylon has been a battle ground. We arrived back at FOB Muleskinner just after day break. There are only a few regular mission days remaining, which should be a cake walk, and then it's QRF only until we leave.

Everyone is packing for home. I put two bags on the MIL van today. Tomorrow we have a customs inspection for our personal items. This is a very detailed and strict inspection of our personal items to verify we aren't trying to bring unauthorized items back home. It feels great to pack for home; I'm now living on bare essentials. I was able to chat with my wife today, and as always it proved to be a mental recharge. I can't wait to see my family. Time for chow and then off to work.

The mission went well. I want as much flight experience as I can get so I volunteer to fly as a co-pilot on maintenance test flights and as many mission flights as possible. I know our aircraft will go into reset when we return and flight time will be a commodity.

16 March 2004

Today was another refreshingly uneventful day. We cleaned an aircraft top to bottom. There must have been fifty pounds of dirt in it. Oily aircraft in a dusty environment equals days of cleaning. Twenty-four aircraft in all means we have our work cut out for us. I finished my day in a great mood after having chatted with my wife for a bit.

At this stage in the war, the aircrews are meeting very little direct enemy contact in Baghdad. The presence of a Kiowa Scout Weapons Team or SWT overhead usually calms the enemy very quickly because they will rarely escape the consequences of their actions. There are specific areas we are forbidden to fly, such as Sadr City. Since my arrival in February, Sadr City has been a no fly zone because metaphorically speaking, it's equivalent to hitting a bee hive with a stick.

Our goal has been to restore order to Iraq and in the eyes of the theater commander, flying into Sadr City is counterproductive. This is a controversial topic in all audiences but with such little experience, I follow orders and maintain a neutral position on the matter.

17 March 2004

I've slowly shifted my sleep schedule to accommodate working nights. I was able to call home again today and talked with my daughter Madison; my oldest daughter had already left for school. My wife told me that Madison was struggling in school with learning to count money. How do you inspire motivation in your child? Better yet, how do you do it from Iraq? I feel so helpless in this respect. All I can do is hope her momma can help her and I'll support them the best I can over the phone. My wife seemed a little stressed and was snappy with me. She's feeling the pressures of being a single parent. She needed a break, something so simple, but I couldn't give it to her. This is such a helpless feeling. Some quality time alone together will be welcomed by both of us.

My platoon leader asked me to work on my OER support form. The support form is a feeder document for each officer's evaluation report or OER. It all seems like bureaucratic crap to me. My focus is on completing each flight mission safely. If you need a few bullet points, I can jot something down I guess. #1 – Stay alive #2 – Kill some bad guys.....as I laughed, I said this is probably not what he's looking for. He may not appreciate the humor. Every situation has room for sarcasm. When the tension is thick, it's an easy way to vent your frustrations.

18 March 2004

It's Thursday and all is well. I am working QRF tonight so I slept most of the morning. I got up in time for lunch chow, and then walked back to my room to see what was left to pack. I chatted with my wife online and then called home to talk with the girls. They seemed okay but my wife said my absence was beginning to hit Madison hard. Hearing this caused a wave of pain to circulate through me. I know I'm needed in two places at once and instinctively family should come first. It does no good to dwell on it so I quickly changed my train of thought and went into mission mode.

Our mission was to patrol assigned routes within the city for any suspicious activity and serve as a deterrent to potential threats that could be brewing in the night. The mission went well and we had no surprises - just how I like it. I flew with a lieutenant from Palehorse troop and one of their pilots said "Make sure he doesn't leave his M-4 out on the weapons pylon." Earlier in the rotation he mistakenly left his M-4 hanging from the weapons pylon on the left side of the aircraft and subsequently departed with it barely hanging on. It was noticed almost immediately so the pilot had to make a very careful approach and landing to recover his rifle. You never seem to outlive stories like this and it would continue to be a point of ribbing for him even after our return home. We finished the mission just before sunrise, and just in time for breakfast. The breakfast meal was a favorite for me. Halliburton was the target of much criticism but I was thankful for the quality of food provided.

19 March 2004

It's Friday! While that means something at home, it means nothing here; it's just another day. I slept in until around 1300 hours, and then went to the finance office to get some cash. I bought a good selection of bootleg movies, not only to enjoy here but some for the girls back home as well. Today is the one year anniversary of our initial invasion into Iraq and could mean an uprising from the old regime.

My work shift tonight was QRF only. It was surprisingly quiet in our area so we weren't launched. There's always a little hidden anxiety in the back of your mind when you're on the hook for a call. One thing is for sure, if we were launched, something had already happened and we would soon be right in the middle of it.

20 March 2004

I'm working twelve hours tonight and flying about five of it. I enjoy the missions and the flight time is adding up. Sometimes it gets a little boring but it has to be done. For me, it's good experience. We have had many attacks across the city in the past few days as this is the anniversary of the start of the war. I purchased some flags to give away back home and will fly them over Baghdad so they'll be genuine keepsakes. I think back to when I first rolled into Baghdad. It wasn't a pretty site and I had a few close calls. I thank God for His protection and the safety that He has provided us. As I hear more explosions and small arms fire across the city today, I pray that our soldiers are safe and return home safely. We fight for what we believe in and that is the freedom of all men. I say "Thank you Lord for your protection," and often quote the 23rd Psalm. "Yea though I walk thru the valley of the shadow of death, I will fear no evil. Thy rod and thy staff, they comfort me." These verses keep me moving.

21 March 2004

This morning, I flew the last four and a half hours of my twelve hour shift. I neatly folded and carried five American Flags with me on the mission. I'll carry those flags home and give them to a handful of my major supporters. My dad will get one, another will go to my closest friend. I will give one to my home church and one for my other friends, Dr. Jack Rushin and Greg West.

The mission was successful last night and I thank the Lord for it. After the mission, we came in and grabbed some chow; then it was off to bed. I did chat with my wife for just a few minutes but cut it short since I was tired and had to get some sleep.

I just woke up from that sleep and I'm well rested for another mission tonight. It should be a shorter flight unless we get a change of mission or a follow on mission, which is not uncommon.

22 March 2004

I flew a mission in the evening, patrolling the same routes and watching the same intersections. We flew downtown into the "Green Zone" which is fun and breaks up the monotony of what we referred to as the triangle of death, not because of its danger but because of the amount of time we killed flying it. The Green Zone was a maze of sky-scrapers and a downtown area that has been cordoned off from the general population to house people that are crucial to the country or military and a few others, such as national reporters. Although heavily guarded, it was often the target of large attacks.

23 March 2004

I was up at 0800 hours and off to chow. After some scrambled eggs and bacon, I packed some stuff for the trip down to Kuwait. After dinner chow, I went to a meeting at 2000 hours where we discussed the aircraft movement plan. Shortly after our meeting, our FOB was attacked by several mortar rounds. Some hit the far western side of the FOB and others landed outside the perimeter, injuring no one. Each attack gets my heart racing while I wonder if one of them has my name on it. The earlier I leave here, the better. I'm working late tonight so I'll eat midnight chow. This late shift is the most challenging because our bodies work similar to an alkaline battery and after 10-12 hours, we tank unless fueled by adrenaline.

24 March 2004

We completed our mission this morning and found one crater from the mortar attack last night. It was an absolute miracle. The rounds that landed inside our FOB did not detonate. I thank God for His protection over us. We flew into the morning sunrise and captured some good pictures. We flew another flag over Baghdad today and will take it home for one of my supporters. The air was filled with smog composed of black soot. Black smoke fell over the city like a cloud of darkness over Gotham. It was nasty and carried the smell of burning oil and feces. I can only imagine the health issues this may cause later. Our squadron commander hosted a unit run on the FOB so we provided aerial security and snapped a few photos of the event from 200 feet above. It's been a long morning; I'm exhausted and need rest.

I woke up around 1500 hours and was well rested. I was able to chat with my wife which proved to be the best part of my day. I was off for a day and enjoyed the down time.

25 March 2004

I ate midnight chow and I'm still up trying to get in contact with my wife, but I think she took the kids to their karate lesson. We watched the Blue Collar Comedy tour at the chow hall and had a great laugh before returning to my bunk for a nap.

I woke up around 1130 hours and headed to the chow hall for lunch. The commander tasked us to clean aircraft from 1300 hours until complete. Nobody enjoyed cleaning aircraft but it meant we were headed home soon. Spirits were high and we tackled the project with enthusiasm.

26 March 2004

I went to the PX and browsed to kill some time. I received an email from Greg West back home and he requested I pick up some coins for his son's coin collection. I have some bills to give him already so I'll get some coins tomorrow.

Tonight I flew my last scheduled mission in Baghdad. The mission went well as we conducted reconnaissance and security operations in the heart of the city. I thank God for His protection over me and our squadron.

This picture was taken after the last flight mission prior to departing Baghdad. Pictured left to right is CPT Hays, CW2 Crandall, CW2Mills, and CW2 Geisler

27 March 2004

Today is my mom's birthday and she's spending it with dad doing something she loves - camping. She's been working a lot lately so she deserves the time away. I packed most of my remaining items today and then washed aircraft for the 3rd day in a row. Other than those couple things, I took it easy and rested when I could.

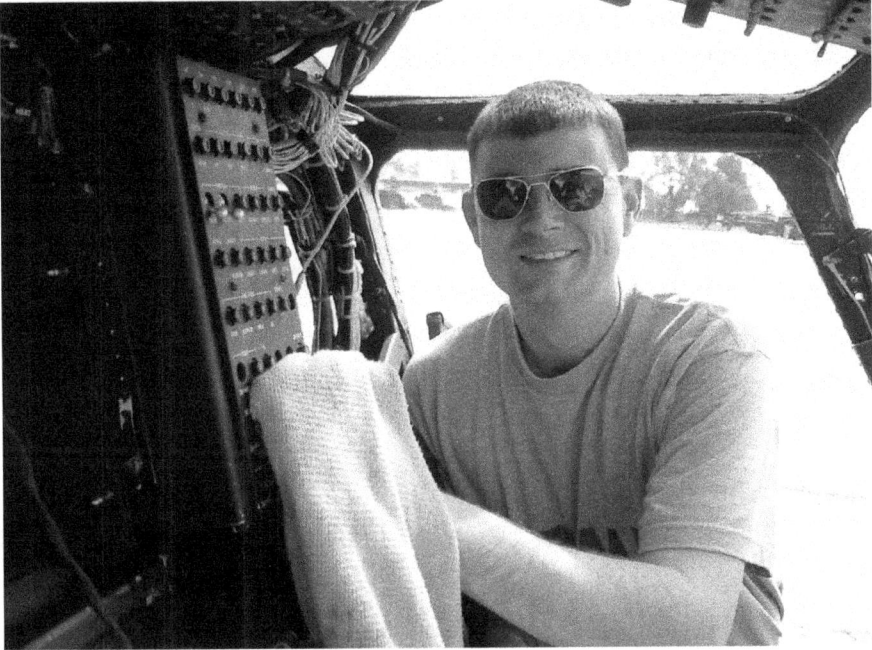

Aircraft cleaning in preparation for returning home

28 March 2004

I woke around 0400 and went to chow at 0500. My calling card had several minutes left on it so I called home to burn up the remaining minutes before leaving Muleskinner. Roberta sounded good and I really enjoyed our chat. It was early for her but she's always understanding and flexed to meet the needs of my schedule.

By day break, the airfield was littered with refreshed and energetic pilots ready to go home. Gear hanging from weapons pylons, missile racks, and rocket pods, we began pre-flighting our birds for departure. We strapped into our Kiowa Warriors, rapidly went through our checklists, and hit our start switches. Engines roared and jet fumes melted the air as we lifted from the pads for the last time. Spirits were high when we left Baghdad behind us. We departed in teams of two staggered by only a few minutes and no one looked back. It was a four and a half hour flight of smiles, sarcasm, and jovial radio chatter. I felt a relief as we crossed the border into Kuwait. The physical tension melted away as we transitioned across the Iraq / Kuwait border. We're making progress with getting home to our families, but the days that lie ahead were sure to test the strength and sanity of not only us as soldiers but also our families.

Chapter 6

Plot Twist

Upon arrival in Udairi, we de-armed our birds, tied 'em down, and unloaded our personal gear. We walked a very dusty quarter mile from the airfield to our assigned tents and dropped our gear. I was able to call home and let everyone know of my wellbeing, and then laid down for a late nap. MWR had a tent in Kuwait that offered TV, games, movies, books, and more. MWR is an acronym for Morale, Welfare, and Recreation and is an organization that supports military at home and abroad. I watched part of a movie there and then supper hit me hard. Needless to say, I had to make a latrine run and I do mean run! Yea, it was one of those that hits you hard so you better get there quick. A port-a-john never looked so good. I'll probably just rest tomorrow. We have to take the birds to the wash racks on Tuesday and get ready for our final Customs inspection. It'll be nice to have that checked off my list and just rest until my flight out of this sand box. The dirt never stops blowing here. Wind and powder dirt equals grit in every orifice of your body, no fun.

I've spent a lot of time in thought about my family during this separation. I believe it has made us stronger but I always look forward to getting home and embracing each of my girls. I love them dearly and realize it more each day. I thank God for giving me an awesome family. I love you girls.

29 March 2004

Today was our first full day in Kuwait and it felt great not worrying about incoming mortars and rockets or if one of the local nationals was a sleeper trying to kill Americans. We lounged a lot and did very little. I called Roberta at 0800 central time but the DSN phone would not allow me to dial her cell number. I called home knowing there would be no answer and left a message to cheer her up. I'm missing the girls even more each day now that there is light at the end of the tunnel. It seems that it's really close even though we're two weeks from stepping on US soil. Tomorrow we'll fly to the wash racks and complete the last cleaning cycle on all the helos before the final customs inspection. It shouldn't take long since we did the bulk of the cleaning before leaving Iraq.

As I sit on my bunk listening to some country music, I dream of the day I'll set foot back home and hug my wife and kids; they really keep me going. I can tell that Roberta stresses a lot, hoping she made the right choices for the family while I've been away. When I'm gone, she has to take care of things the way she sees fit. People make different choices but hers have been great and I'm proud of her for doing so well. I know I'm hard to please in the decision department but she has done great. I looked at gold necklaces for her at one of the local shops. There are a few that would look great on her at the Squadron Ball but they are expensive. Nothing is too good for her but what she deserves usually exceeds my budget. I love her and would have no one else.

30 March 2004

This morning we flew from Udairi to the wash rack area in Doha. It seemed like we washed the birds until the paint was thinning. As a new pilot to the unit, I was placed with a seasoned instructor pilot to carefully guide me through the washing process so I didn't inadvertently break or damage anything by spraying water in the "wrong" places. I attentively listened to a ten minute speech about how to carefully move about the aircraft and what not to spray. He finished his dissertation with specific instructions on walking underneath the tail boom and how I should pay particular attention not to hit a small antenna that extended below it because it was easy to break inadvertently. Just after finishing the speech, he crossed under the tail boom and WAP! His head struck that very antenna and broke it smooth off! I laughed a bit with a smirk and shook my head as we continued.

We spent about three and a half hours washing aircraft before the customs agents showed for the final inspection. We passed with no trouble, thank God. We finished and were bused to our holding area which was crowded with more soldiers waiting to return home. Tomorrow we'll fly the birds to port for final preparation and packing before being loaded on the transport ship.

Feeling relieved and anxious to be home, I spent some time jewelry shopping and settled on a set of pearls. It was a medium length strand and included a bracelet. As often as she wears the fake pearls she has, she's sure to love the real ones. She deserves real pearls, diamonds, and gold, even though she rarely wears jewelry.

31 March 2004

We awoke to the sound of second squadron's first sergeant yelling a wakeup call to his soldiers. "Let's go boys!" It felt so early to me. I needed to get up and moving anyway. I dressed and went to chow. I fly at 1000 hours and this should be the last flight before the port operations team receives our aircraft. It'll be nice to pass the aircraft to them and not have a care in the world besides getting back home.

So here I am sitting on my bunk, typing in my journal, listening to Clint Black, and thinking of being home with my lady and little girls. I love you girls. I'll probably spend some time looking around at everything here in Doha today. I can't believe the facilities here. This is like being back on post but... much better! They have everything. There's a library, a surprisingly large post exchange, and an extraordinary gym with basketball, boxing, and volleyball. There's a free movie theatre that plays most of the latest run movies, so I'll be sure to take in a few movies while here. The rumor mill says we may get some early seats back to the States. I don't expect I'll get one since I arrived so late.

1 April 2004

Today is my dad's birthday. I emailed him to say Happy Birthday. I know he's feeling older, but hopefully his spirits are up. I missed Roberta on the computer today so I had to email her. She'll be mad that we missed each other. I bought the girls a couple of little watches as gifts; I know they'll love them. We've done a lot of nothing today which feels odd after continuously be busy for several months. We did play chess at a place called Frosty's something or other. I went to the theater with a fellow pilot to see the third Matrix movie. After the movie, we ate at Subway and just visited for a while at the table. I talked to one of the mall vendors while waiting on the Lt to use the internet. The vendor told us of her travels to Maples and how beautiful it was.

2 April 2004

I was up by 0700 and off to breakfast. The breakfast food isn't too shabby. I went back to my bunk to read a magazine for a bit before taking in yet another movie. I was able to call Roberta today. She sounded good. When we are away from each other for a long time, I realize it's the small things in life that we take for granted. Things change when people are separated. She takes care of things at home and does a great job. Sometimes she doesn't do the same things I would, but that is because everyone is different and that's what gives the world its variety. Variety is good. I think our relationship is getting stronger. I am thankful for that.

3 April 2004

0730 felt like 0500, but I rolled out of bed and went to the chow hall for a hot breakfast. It was nice to sit and drink a hot cup of coffee with plenty of time to spare. Not that this hasn't been available, but with no schedule to keep and no work to be done, it was relaxing. After breakfast, I had an internet chat session with Roberta via webcam. It was great to see her. The kids were anxious to see me and continuously jumped around in excitement.

There's an area in Doha named the Marble Palace that was designated for rest and relaxation for troops on their way home and troops that opted not to return to the States during their fourteen day R&R period. What better way to kill some time than at the Marble Palace? We took a bus around 1100 hours and went to the palace. We were able to hang out by the pool and soak up some rays. I thought it might be nice to put on a bit of a tan before returning. We tried our hand at tennis and just as I thought, it's not my sport. It's great exercise but requires some skill.

The pool was an oasis of ice water, too cold to get in but beautiful. We returned around 1600 and I was able to chat with the kids for a few minutes. After supper chow, we lined up for the movie "Torque." We waited in line for about an hour, not because it was a great movie, but because there's nothing else to do to whittle away at this block of time that separates us from our families. After the movie, I came back to the barracks and closed the day by laying my head to my pillow and falling asleep listening to music.

4 April 2004

My body clock didn't allow me to sleep in so I started my day at 0615. Apparently, others didn't have the same issue so I went to breakfast solo, then spent some time thumbing through a magazine. I imagined being with my wife, enjoying the back patio of our new house. Having deployed shortly after our household goods were delivered, I had little time to enjoy our new house. I'll be with her soon.

After breakfast, I went back to the barracks bay and visited with one of our Lt's about all the things we wanted to do when we got home while we awaited the arrival of baggage from the "main body." The main body is the largest group in the unit movement. They arrived late and it rained, so we delayed the unloading operation until after lunch. We unloaded two mil-vans of baggage before going to supper chow and were eager to find our bunks after the work detail. I ended the night with a game of chess against a fellow Warrant Officer.

Word circulated that they filled the water reservoirs for the showers so I took a quick shower as well. The shower water runs out quickly so you have to catch it at the right time. The showers are gravity fed by large water containers and empty quickly with so many soldiers using them. We were scheduled to spend fourteen days in Kuwait prior to returning home. We're down to ten days remaining and it feels like a lifetime. The time is grows shorter daily.

6 April 2004

Today I've been tagged, along with another Warrant Officer, to be on guard duty. I can't believe I am actually guarding vehicles from being tampered with by our own Army! I was sitting in a Humvee on a twelve hour shift guarding vehicles from being vandalized by other military units. Twelve hours is a long time when there is nothing to do but look at trucks. After a long boring day of guard duty, we came back, worked off some frustration in the gym, and then grabbed some chow. One more week until we head home; time is standing still.

7 April 2004

We decided the pool would be on our activity list today so we headed over there around 1100. We hung around by the pool getting our tan on for an hour or so. We played some air hockey and ping pong to kill some time and then went back to the camp around 1600. We came back for some dinner chow and to our surprise they are serving lobster, T-bones, crab legs, and shrimp. Wow, what a surprise! It wasn't like grilling back home but it was good.

We went to Frosty's game lounge to see what was going on there and they were having a Bingo night so we stayed and tried our luck. The LT won on the first round. I didn't win anything. Afterward, the flight doc said they were giving Salsa dance lessons at the gym so we headed over there, intent on improving our moves for the ladies back home. We had a great time, but I'm glad there were no video cameras. I rounded out the night with an ab workout and then a shower. I hit the sack at 0030.

8 April 2004

We started today like every other boring day, with breakfast, a cup of coffee, and thoughts of how we could make the day go by faster. However, nobody knew what this day had in store for us. We went to the PX and picked up a magazine and sent out some emails. Upon returning to the barracks, we start hearing nightmarish rumors of possibly going back into Iraq.

My stomach began to churn. We are four days and a wake-up from seeing our families and they're considering sending us back to Iraq. I had so many emotions running through me all at once. I had a sinking feeling in the pit of my stomach and began to feel physically ill. I can't believe this is even being considered. My heart started to beat faster and thoughts raced through my head. The rumor mill is in full swing and the troops need reliable information. Tension is understandably very high right now.

I was able to call my wife and let her know what she may hear. We're in a holding pattern on whether we go or stay. It's an emotional roller coaster that tests your mind, will, and emotions. I quickly learned why sometimes it's important not to be prematurely informative to your family. We had no news at nightfall. I pray to God that we go back to our families. I'm willing to do my job if needed but also miss my family tremendously. I sent an email update to my wife. No news is good news for us. There's been an up-rising in Southern Iraq and another unit did get the word to expedite, so they are moving north starting tomorrow. War is a sad thing; I feel bad for anyone who has to endure it. Lord, I pray now that we will rejoin our families as scheduled and please be with those who cannot be with theirs. Amen.

9 April 2004

It's Good Friday but feels more like black Friday. We were just informed that we will in fact be going back into the fight. I'm more hurt for my family's sake than for my own. It's imperative that I appear strong as a leader and as a soldier. I must be strong as an officer, leader, and husband to my wife. Our briefing included information that the area we'll inhabit has little to no facilities, so I picked up a few items from the PX. I also bought a phone card and called home. Roberta sounded heart broken and upset. I couldn't decipher if she was tired or masking her emotions. I was on the verge of tears but held it in. I had an epiphany of the purpose of the countless mental conditioning exercises through every phase of my training. You must be prepared to mentally adapt to any situation thrown at you.

I linked up with Bertie online this evening and learned that her day was as emotional as mine. This really makes me hate war and the differences people have with each other. I'm beginning to understand my father-in-law's outlook on personal property. It means nothing to me right now. I feel ashamed for being so vain about goods. I am changing and growing to value worldly goods less and less. I have always been really protective of my personal belongings and in the past have gotten upset with my own family for little mistakes that are just accidents. It makes me ashamed of myself for being so tough on my kids for little things like that. It seems so insignificant now. Life is too short to be upset with each other over the small stuff. I am going to be more tolerable and loveable when I return. I regret how I treated my wife at times because of my lack of patience. I am sorry for that.

I just finished chatting with Bertie and I'll be calling home tonight. I can't even explain the feelings I felt when they said, "It's true; you're not going home in four days. We are being extended for up to 6 months." I had a sick feeling in my stomach, as I'm sure the girls did too. I feel like I went flush. It was like a bad dream. How do I explain this to my seven and nine year old daughters? They don't understand and fully trust me when I say, "I'll see you in four days sweetie." It broke my heart to hear them cry; I love them so much. At that moment I was so angry. I was mad at the Army. I was mad at the up-rising militia that caused this and I was angry with the situation in general.

Chapter 7

Extension

10 April 2004

It's been a very tough morning. I can't stand the thought of breaking my girls' hearts because I can't be home when I said I would. I'm sure Roberta is crying buckets of tears right now and there's nothing I can do about it. This is one of the most difficult moments of my life. I pray we can get this finished quickly and return. I went to breakfast but only ate a few bites. My appetite is gone. I haven't had an appetite since we received the bad news. Maybe this will prove to be a great weight loss program for me. I would love to surprise her with a great toned body when I get back. Roberta's parents are going to Louisiana today to pick her and the girls up and take them to Missouri for a week. This support should help keep her focused and transition through this tough time.

Since I'm not going home right away, I mailed a package of books and gifts I had for them. I called Roberta this morning and was able to chat online with her. I called my brother Rod and we talked for an hour. He was encouraging as well. God has me here for a reason so I'll have faith and step up to receive this strengthening experience. "Lord, please get me back to my family quickly and without incident." I silently recite the 23 Psalm as I contemplate the battle ahead of us.

> *The Lord is my shepherd I shall not want, He makes me to lie down in green pastures, He leads me beside the still waters, He restores my soul, surely goodness and mercy shall be with me all the days of my life. And yea though I walk through the valley of the shadow of death, I will fear no evil. For thy rod and thy staff, they comfort me.*

11 April 2004

With as much sarcasm as I could muster, I said, "It's another depressing day in Doha paradise." Now as I wake up each day, thoughts of home slip further away. The morning hours are the most depressing, but I'm coping with the realization of facing death once again. I hope my family isn't permanently distraught from this emotional train wreck. My motivation level is at zero and holding. I try not to dwell on things back home but it's impossible. We are nervous and worried about our families. Many people had life-changing plans that hinged on their return home. Marriage, vacations, and even divorce, but it will all have to wait.   Liberty and freedom is worth much more than I ever realized.

12 April 2004

Today was full of activities in preparation for the move back into Iraq. We spent the day unpacking equipment, turning in broken equipment, trading broken gear and vehicles for operational units, all the while knowing we should have been boarding a plane bound for U.S. soil. What a mind trip. The nightmare has seared into my mind long enough that it's now a reality. After seeing the condition of our gear and knowing how the helicopters looked when we put them on the ship, I felt like we were returning crippled with minimal gear. I had an epiphany and realized that in every phase of training prior to deployment, we were given a change of mission, a curve ball. Each of those was designed to condition us for this moment. I was conditioned to take the mission as it presents itself and execute. If you can overcome your own mind, you've won half the battle. We are the United States Army; we'll make it work and complete the mission, no matter what it takes.

13-16 April 2004

I am up early for departure and nervous about the flight back into Iraq. "Lord, please protect us all on this journey. Amen." I keep thinking of how much I miss my family and wishing I were going home to see them. It's not happening today, so I may as well shut that down right now. I just returned from breakfast and picked up my M9 and M4. I'm now carrying two weapons. With my M9 tucked away in a broken make-shift Vietnam era holster and my M4 slung over my shoulder, I headed back to my bunk to pack my go-bag one more time. We're all pretty pissed about not going home so our tolerance for shenanigans from the enemy is minimal.

I'm sitting on my bunk and don't know what to feel, but I'm feeling several emotions. I'm mad, sad, scared, a bit unhappy, and very sorry that I hurt my wife and kids by inadvertently misleading them. I can't control what happens here so it's not my fault but I feel responsible. I'm here to do a job and I'll do what I swore in to do.

I'm nervous about what we're getting into. Our briefing revealed that we would split up and operate from two different FOBs, Babylon and Al-Kut. My troop was assigned to Al-Kut, where the Mahdi militia had overthrown the Ukrainians at their downtown outpost only a few days ago, killing several coalition soldiers. "Lord, I ask while I'm committed to this mission that you protect all of us and bring us home safely to our families, thanks in advance."

After our briefings with the S2 intelligence team and the commander, we moved to the flight line and then to the helos. We weren't sure what we may be getting into upon arrival at our FOBs so we were loaded for bear. We had a full complement of ammo on us and would load the birds with rockets and 50 cal just before crossing into Iraq. As a low time pilot, I would be in the last chalk of the flight, chalk four.

The commander positioned the experienced crews in the front of the charge. Our troop departed in two flights of four chalks each, a total of eight aircraft. I was paired with a maintenance pilot and we were flying a crippled aircraft. Under normal circumstances, we wouldn't think of attempting a flight in this aircraft but it had to come with us. It was safe enough so we executed.

The engine and transmission torque gauge was completely inoperative and some of the radios weren't working. Total flight time jumping back into the combat zone was about three and a half hours. We flew from Kuwait to Talil air base where we armed the birds for the final leg of the flight into Al-Kut. Our arrival into Al-Kut seemed anti-climactic. The city was calm, thank God. Not a shot was fired.

Al-Kut is a location with very few amenities. We moved into a hanger that was full of bird poop, had no power, and every window was broken. After the Iraqi army abandoned the airfield, the locals looted the building for anything it had that was worthwhile. They took everything. The wiring was stripped, doors removed, even the light fixtures were gone. I embraced the suck of the situation and stayed sane by laughing. I kept notes on paper for the days we didn't have power. We're operating by flashlight at night. The Tactical Operations Center is on generator power. We ordered some industrial sized generators to power our area on the FOB but that will take at least a week, if they're even available. There's only four port-a-johns in our area and they filled up the first day.

Shortly after arrival in Al-Kut, trying my best to smile.

We have flown every day since jumping back into the combat zone which is good because it chips away at our open ended countdown to going home. At this point, we only know that the extension may be as long as six more months. That would mean enduring the summer's sweltering heat. I am getting some good flight time while here.

The event that triggered our return to Iraq was a large fire-fight that ended with us killing more than 300. I say we but it was our ground troops that did the trigger pulling. Since we've arrived, the city has calmed so we spend hours each day patrolling routes and providing perimeter security for the FOB and other coalition checkpoints throughout our A.O. We had a team out flying that found a weapons cache. They requested permission to destroy it but our higher command could not get clearance of fires from our parent unit before running low on fuel so they returned to base without firing. Our fear is that the weapons will be picked up and used against us.

We jumped back into Iraq in such a hurry that the FOB in Al-Kut wasn't logistically equipped to support our unit detachment. They could only support feeding us one hot meal each day, dinner. It was the highlight of my day. The food was good but the line to get into the chow hall was a torturous forty-five minute wait. It had air conditioning, good food, and hot coffee so we leisurely had dinner and enjoyed the cool air. We had a small haboob today which is like a dust storm. It was just bad enough to make the dust linger in the air for the day. The main body rolled in today and began setting up the command center. Al-Kut had a phone trailer that was serviced by satellite. We could purchase phone cards to call home. I did so and was able to talk to the girls using a phone card. I miss them greatly. Kuwait seems like a dream that never even happened.

17 April 2004

It's Saturday and that matters none other than we're one day closer to getting home. Everyone is unpacking gear but the plans keep changing so there's no telling where we will end up. We're hearing rumors of a possible move after a short time in Al-Kut. I called home last night so I didn't even try tonight. The wait for a phone was an hour and a half. A long wait but well worth it.

We ran an extension cord into our room tonight so I was able to use my laptop and write in my journal while charging the battery. The lines at the chow hall have been so long that they've been bringing chow to us for each meal, which is really nice. A small group of the pilots huddled around a bunk and played cards while darkness slowly coated the night sky.

Many things have changed about me in the past few months. I think they're all for the good. I have enjoyed reading more. We'll see what Roberta thinks. I hope I have a high tolerance level for the kids when I return. We're told that patience won't best our best quality.

The nights are cooler so the mosquitoes come out in full force. They haven't been too bad for me because I use a lotion that has bug repellent in it. It's the best I've seen so far. We're living in tight quarters in an open bay type area. My domain is approximately five feet wide and seven feet long, just enough space for a cot, my ruck sack, some boots, and a small table for my PC. I have food, a place to sleep, and power. Life is good. Well..... better than a few days ago.

18 April 2004

I should be going to church with my family today but I'm stuck in the desert fighting the fight for freedom, for a people that has never known freedom. What will they do with their freedom? Will they appreciate that I'm gone from my family to help their cause? Will they appreciate the 18 year old soldiers that fought and died while freeing their country? Will they thank God for a young soldier that gave his life at such a young age, not only for someone they did not know but merely for an idea that he believed in? I miss my family this morning. I think of being home often. I need the flight time but not as much as I desire to be home with my family. I want to do my part for the country but it's hard right now because I thought I'd be home. It's very difficult for me to let go of the idea that I was five days from being home and they extended us. I'm trusting God to take care of me and bring me back safely.

We are continuing our effort to improve our living accommodations. I'd love to see a shower trailer at some point. As it stands, we're only able to clean up using baby wipes and pan water. I went to church this morning. It was the first service here at the new location. We received news that one of the convoys made a wrong turn and was ambushed, killing 3 soldiers. My heart sank. Today could be my day or maybe tomorrow. I can't say it enough times, when does the killing stop?

The hangar we're living in needs some real TLC. The windows are broken out and glass, dirt, and bird poop is everywhere. Some locals were hired to come and clean up the hangar area so we could remain focused on the mission. The only trouble is we don't know who to trust and there's no one here to watch these guys but us. We're back in Iraq because of an up-rise of a local militia so I don't know who to trust and have to expect that everyone could be trying to kill me.

There were a total of about ten workers they let in to do the clean-up in my area. I wanted to make sure they were very clear about my intentions. I was over-seeing the work of five of them.

There was an obvious language barrier so we were limited to hand motions, eye contact, and tone of voice. As they approached, I pulled a full thirty round magazine from my vest, inserted into the magazine well, and proceeded to chamber a round, making certain that they understood my point of view. Five to one isn't good odds so I kept my distance and didn't lower my guard. I had them picking up broken glass and doing some general house cleaning type stuff when one of them began saying something and pointing at the ground. I motioned for him to clean and he said something again, taking steps toward me. I took a step back, made eye contact with him while I readied my weapon at a forty-five degree angle, and motioned for him to back away. He did and continued to pick up broken glass. I could tell he was irritated but I'll be damned if I'm taking any unnecessary chances. They finished the clean-up and I escorted them back to their pickup point so they could move on to their next task.

The routine pilot schedule was published today and I'm on the day shift so that will be a bit different since I've been used to flying nights all the time. Day shift will be a little more fun. I do love the low and fast tactical flying. It's intense to be at fifty feet and 100 miles per hour. What an adrenaline rush!

I called home today and talked to Bertie. She seems to be doing well. Her dad was mowing the lawn for her. He has helped us a lot during this hard time. He knows what it's like on both sides, stateside and deployed. I appreciate his help now and his contribution during the Vietnam War as well. He had it rough dealing with PTSD from Vietnam. He was Navy man and commanded a brown water patrol boat in the rivers of Vietnam. War sucks! There is no other way to put it.

19 April 2004

Today we moved our bunks to different rooms just trying to get everything situated. I flew two missions that totalled four and a half hours. Our mission set had us patrolling roads and looking for munitions that were remaining from the Saddam regime. Each major city had bunker positions around it. Each of those positions had some type of ammunition in it, mainly mortar and tank rounds. Our job was to report those so our EOD teams could round them up and destroy them before our enemies could make roadside bombs with them. I'm thankful for the flight time and experience. I was quite busy today so time flew by. That's the way I like it.

I'm crewed with CPT Hayes so we talked while flying today and we're in agreement that we wouldn't be able to discipline the kids for some time after our return. I miss them so much; I'll probably spoil them for a while. We see the local kids here and feel sorry for them. They have little to begin with and most are poorly clothed. I want to drop candy to them if I can ever get into the PX to buy it. The line to get into the PX is two to three hours. Tomorrow my flight team is off so I may spend that time getting into the PX. I need to get some clothes out of another mil-van and need some laundry soap for washing. My washing will be done by hand for a while. They're working on installing electric in the building right now. Hopefully that will be complete soon.

I saw several foxes and huge jack rabbits in the desert today. The airfield is separated from the main populous and can be a bit spooky at night. I hear the owls come into the hanger in the dark hours and eat the pigeons. It's eerie to listen to. They screech and the slightest noise will echo through the metal hangar like special effects in a horror movie. Cultures are very different here as well; we see many families living in tents in the wide-open desert. They're remote from everything. We see many shepherds as we're flying the area. They tend their sheep day and night. Even in Baghdad, there were shepherds that tend their flocks in and around the city. It's different. It feels like I stepped into a Bible story. Much of the biblical history took place in Iraq. Our sister troop is basing in Babylon, land of a once great empire.

20 April 2004

Today I'm trying to get some administrative things completed. I need to go to the PX and pick up some laundry soap to wash clothes. I haven't washed any clothes since we left Kuwait six days ago. I try to get several days out of each set of clothes and minimize the need to wash. I've not yet taken a shower since leaving Kuwait. Baby wipes have been the best alternate and they leave you with a tacky film. They're good to have but nothing like the great feeling after a hot shower. There are some showers down the way that belong to another unit but there's no hot water. It's better than nothing, so we're going to attempt a late night run in hopes we aren't seen by anyone. Water is a commodity and well protected so it'll have to be a black op mission.

Captain Hays has a load of country music on his PC that I'd like to get from him. Listening to it reminds me of home and can be painful but also serves as a mental escape. Ironically, the thing that gives me a short mental vacation also causes me emotional stress. I pray the Lord helps me grow from the hard times. I know my family and friends appreciate my contribution to our country but I'm not sure I'll choose to do this as a career for twenty years. I'll evaluate my options when my six year commitment is complete. For now, I'm saving lives and building flight time. I looked at some pictures from back home just before bed and it really made me homesick. I can't do that many more times. I need to put it out of mind and stay focused on the mission.

21 April 2004

It's 0800 and I'm up sipping some coffee while I plan my day. Last night during the troop update brief, the commander said they had contacted some pilots that had already made it home and they'll be rejoining us soon. I'm sure they're thrilled that they'll be coming back. Flight time here is handed out like candy while back home we fight for it. I'm doing some database development that will aid us in record administration back home. It's a time consuming task that occupies my mind and kills time.

Our unit installed an MWR phone so I was able to call home for free from the hangar. Wow, what a morale booster! The MWR phones are operated by satellite. Now that our command center is up and running, we have the ability to use internet as well. It's extremely slow but it works. I usually write emails offline and then connect to the wireless just long enough to send it. Even that takes about twenty minutes.

I wrote a letter home today apologizing for past actions and such. I feel bad for the way I have acted in the past by letting accidents that occurred drive me to being overbearing and harsh to my wife and kids. I feel I may have been hard to live with from time to time. Some of that comes from my being so detailed about caring for material things. I don't want to be that way. I'm sure I'll be very different when I get home.

22 April 2004

I slept in until 0830 then got up and shaved like usual, with my portable mirror and some bottled water in a canteen cup. It's quiet in the hangar except for a few birds fluttering their wings and the slosh of my razor being cleaned in the canteen cup. I'll feel like a king when I return and have hot running water anytime I want it, a shower without wearing flip flops, and clean clothes every day. Sometimes we play the hypothetical game. "If the army offered you a check for $10,000 for another 3 months in Iraq or a plane ticket home what would you take?" Most everyone says "I'll take the plane ticket, the money isn't worth it". Nearly everyone feels the same way now. I feel that way and have only been here a few months and these guys have been around for a year so I'm sure they're tired of it.

We had a big dust storm today with winds up to 50 knots. It was a mess around here. Needless to say, we didn't fly our mission. I'm scheduled to fly with the squadron commander tomorrow so we'll see how the weather is by then. I killed some of the day by watching a movie. I have about 54 movies so if I watch one every other day, by the time I've seen them all it will be time to go home. That would be 108 days, and since I started late, it should work. It's time to get some rest for tomorrow's flight. I'm flying with our squadron commander so I better be on my game.

23-27 April 2004

The days are getting much warmer. It's been near 100F each day. The missions are becoming more demanding. Our team flew six hours today. During a route reconnaissance mission, we were tasked to provide aerial security for a convoy that had an accident on a canal bridge. The driver of a Humvee struck the back of another vehicle which caused the rear of the vehicle to shift left abruptly and land his left rear wheel off the edge of the bridge.

The soldier was not wearing his seatbelt and the jolt threw him out of the Humvee into the canal. He was in a full ensemble of body armor and was far too heavy to swim. This was not a rescue mission; sadly, it was a recovery mission. We provided security and assisted in the search. It was a bad deal all the way around. We were unable to find the soldier's body so a scuba team was called in from up North somewhere. The scuba team arrived late in the evening and would recover the soldier the following day.

While we were searching for one of our own, our sister troop, based in Babylon, met direct enemy contact in Najaf and quickly found themselves engaged in a fire fight with the insurgency. I didn't get all the details, but one of my fellow flight school pilots had a confirmed kill with his M4 from the cockpit. I was jealous but also thankful it wasn't me. I was extremely tired after the long mission today. Upon our return to base, I crashed out on my bunk and didn't even clean up. Mike, one of the returning pilots, stopped at our base before he departed for Babylon to link up with Nomad Troop. I was able to visit for a few short minutes and then he left. His time at home was short lived and he was sent back for the extension. He seemed ok but we only spoke briefly.

28 April 2004

I started my day by watching a movie. I received some mail from home, the highlight of my day! Others have been receiving back logged mail for a few days so I knew it wouldn't be long. Adrian, one or our CW3s, received a cake from his mom so we all shared it. It was a nut cake or something. I really didn't care what it was, anything different was good.

I loved the letter from my daughter, Naelyn. She is a sweetheart. They sent me a new picture of Madison, too. She's really growing up and looks so much bigger now. I miss them both tremendously and hope we go home soon. I love reading Roberta's letters. She's doing a great job taking care of stuff back home. I'm proud of her. Although we're getting stronger thru this experience, I don't really enjoy the growth, only the results.

29 April 2004

I got up at 0700 to be ready for my 1000 flight. Our mission was to provide security for a ground convoy and provide area security for the convoy, the FOB, and the city of Al-Kut. Most of the people here seem to be glad we are here. Some won't respond to our presence at all while others respond negatively by shaking their fists at us or just waving their hands in a manner to "shew" us off. The shepherds in the desert don't care too much for us. Most likely they get irritated at us scaring their herd. The southern part of Iraq is riddled with quite a few shepherds and herds. The herds are usually either sheep or camels. They occupy the city, rural areas, and roam in the middle of the desert. It's not uncommon to see sheep or goats on top of houses in the city.

During our security mission, we flew over a house outside the city and the roof was on fire. The roof was made of straw and the house built of mud. After quickly deliberating within the crew, the Lt said he wanted to help so it was decided the lead ship would land while the trail ship provided over-watch. The lead ship picked a landing zone up wind of the fire and landed. Our Lieutenant quickly hopped out of the aircraft and hastily ran to the house, fire extinguisher in hand. There was a man on a ladder throwing what water he could find but it was doing little to stop it. Our LT pulled the pin on the fire extinguisher and handed it to the man. In great anticipation the man squeezed the handle and .... nothing happened! The fire extinguisher didn't work! Embarrassed and without saying a word, he took the broken fire extinguisher and returned to his aircraft. The man tore off the remaining good straw so there was nothing left to burn. We watched it all from the trail aircraft while orbiting three hundred feet above. As bad as the situation was, I couldn't help but laugh. At that moment, he represented our country. They didn't see some guy coming to help. They saw the United States Army. Our LT had great intentions but it ended in embarrassment and disappointment. This was sure to become a great story within our Troop.

30 April 2004

I spent my day resting and working on my software project. I was on standby for a mission but we never received the "go" call. The mission was scrapped because of odd circumstances. I did call my wife and enjoyed the break.

1 May 2004

Bertie said she would send my wireless network card soon. We now have a very basic wireless network setup. It's just enough bandwidth to connect and use email. I went to bed early and got up early for breakfast. It was good. I love to start the day with a good breakfast but since working nights, I'm not used to getting up at 0500.

We didn't fly today due to high winds. It was a much appreciated break. The weather wasn't terrible but the wind gusts were above acceptable levels. We had an awards ceremony and a promotion ceremony. A commissioned officer along with a warrant officer both pinned their new rank.

Several air medals were awarded for service. Each unit awards their medals a different way. 4/2 ACR awarded each aviator with an air medal for each 6 months of combat flight. This is very different from the Vietnam War where I understand that helo pilots were getting one for every fifty hours of combat flight time. I didn't care as long as I made it home safe and saved a few lives in the process.

2 May 2004

I slept in until 1000 hours and missed church. I intended to get up early and go to breakfast and then church but I didn't hear my alarm. The rest was much needed. I had a few things I wanted to get done but the day didn't go as planned and after lunch, I helped one of our maintenance test pilots by flying with him during a test flight. I called home and talk to the girls. The little ones are doing well. Madison is doing so much better in school and I wonder if it's because I'm not there stressing her out about grades. Poor grades were always followed by stern speeches. I knew she could do better. She is doing well now. Roberta says she's not worried about her anymore.

I'm not on mission today so I can relax today. I went with some other guys over to the Regimental Support Squadron on the other side of the airfield to watch the NASCAR race but we were 3 hours early. It didn't start until 2300 hours our time which is too late for me since I'm working tomorrow. I killed some time playing Airport Tycoon and that was about it before calling it a day.

3 May 2004

I am up about 0830 and drinking my coffee. I'll go check my email soon if the internet is working. It's intermittent at times, which gets to be a real pain. I'm sitting here writing this and listening to country music. The music is great but also makes me miss home. I enjoy going to the chow tent because they play country music most of the time. It's nice to sit back and sip some coffee while listening to music; it sooths my nerves.

Back home my dad and I used to frequent a place called the Donut House. It was owned by a very close friend of ours and I think of the day that I'll sit in the Donut House with my dad and talk this stuff out. Mike lost his father not long after we arrived in Iraq. He went home for the funeral and then returned for the extension. I can't imagine losing my dad right now. I have so many things I want to do with him. We still have a fishing trip planned. I so enjoy our fishing time. We always have a great time, even if the fish aren't biting. I recall a trip when I ironically hooked myself in the lip while trying to bite the line into and was mad because we had to go to the hospital to get it removed. We had only been at the lake a short time so I considered leaving the hook until we had fished the day away. That was a Saturday I'll never forget. Another time, we were fishing and a keeper fish jumped in the boat, strange but true. We always have a good time.

I spent most of my day playing Airport Tycoon. It's a little addictive. I had to fly at 1900 hours so I just rested until my flight. I flew with CW4 Duerst, one of our more senior pilots. We get along well. We both share an interest in NASCAR. We completed a routine night reconnaissance mission and reset our currency clock for night vision goggles. This is something that's required to maintain proficiency using the NVG's.

Upon returning to the FOB, I was told I would accompany the Squadron Commander and Captain Hennigan to Babylon this week. As a young officer in the unit, I was tasked with the things the senior warrants didn't want to do, like fly with the Colonel. I don't mind the Squadron Commander at all. He was a nice guy and I enjoyed his company.

I was busy flying this evening and didn't get to dinner chow but managed to make it to midnight chow. It wasn't that great. I skipped the dried up leftover meat and went for the starch. I also had a bowl of cereal. If all else fails, the cereal is good. I concluded my day with a shower.

4 May 2004

Today was a bit of a rest for me. I wasn't scheduled to fly so I spent time in the air conditioning while working on the tactical flight planning computer. There seemed to be some issues with the software so I was attempting to make it usable again. Our unit purchased a satellite phone to use for calling home after having many issues with the reliability of the MWR phone. I used the sat phone and talked to the girls for a few minutes. It was quite refreshing. Bertie's doing an excellent job taking care of things. She asked me about the credit card bill for her pearl necklace. Oops! She wasn't supposed to see that. I expected her to be wearing it long before the bill arrived. That didn't work out as planned. I'm looking forward to taking her to the Squadron Ball when we get back. I promised them a trip to the lake when I get home. I'd like to rent a boat and take them out for the day. Now I'm day dreaming. I better go to bed early because I have a 0400 flight.

5 May 2004

I was up early for my flight. I enjoyed flying in the cool morning weather. It was uneventful and we were able to see a beautiful sunrise. I had to prepare a bunch of info to brief the Colonel for tomorrow's flight. When troop pilots fly in a team with staff pilots, the troop pilots conduct the briefing. I am the only line troop guy in the flight so I'll be conducting the briefing. I'm a little intimidated by the process but it will be good experience for me.

Our maintenance team needed an aircraft ran up so I offered to help. I'm always looking for more ways to learn and be a better pilot. I did the maintenance check and all was good so we have one more bird ready for flight. I'm gaining more experience by the day. After shutdown, I went to chow and tried getting on the web but the connection was down. I'll try tomorrow again over in Babylon. Hopefully, I can link up with Mike and see how he has been. I totaled up my flight hours and I have 332 total hours so far. I have over 100 hours of night vision goggle time, not bad for a green pilot like myself. Back home, it would take two years to accumulate that flight time.

Chapter 8

Ancient City of Babylon

6 May 2004

Today I'll fly to Babylon with our Squadron Commander. Nomad Troop is another OH58D Kiowa Warrior troop based in Babylon to support that operations area. The commander wants to spend a few days flying missions with them. I've been nervous about today because I was slated to conduct the air mission briefing for our flight from Al-Kut to Babylon. I have been involved with few formal briefings at this level and it was only yesterday that a senior warrant officer was humiliated by the colonel for his poor briefing technique. I sat in on the brief to gain some experience and learned what not to do. I hate making a bad impression so I did a little homework before the briefing this morning and it paid off. I briefed the colonel and everything went well. He commented on what a good job I had done which set a good tone for our flight.

During the flight over, we conducted a route recon mission over the primary supply routes looking for IEDs and anything peculiar or out of place. We made it to Babylon without issue and encountered nothing to report. The first thing I noticed was that Babylon is full of palm trees. It's a neat city. There's a palace on top of a hill near the city center that displays itself well. We could see it clearly from quite a distance.

Palace in Babylon

Upon arrival, we heard that an Apache helicopter had been hit with a rocket propelled grenade (RPG) and another that sustained small arms fire when they flew near a mosque in Ah-Najaf. The RPG detonated on impact and many of the onboard systems failed. The crew was able to limp back to base and landed without any casualties. The second Apache in the team received small arms fire and three of the rounds went thru the cockpit but all missed the pilots. One round came into the gunner's window and over his head, ricocheting down and out the other window, a second round came thru the pilot's side window, ricocheted off of his helmet, and went out the other window. His angels were pulling overtime. The bird hit by the RPG was badly damaged but made it back. After returning, it was approved for a one time maintenance flight to a major repair depot. I missed seeing it before it left but reports said it looked bad.

In the evening, I went over and took pictures of the Babylonian ruins and pondered the history that had occurred exactly where I was standing. Some locals have been digging into the ruins and discovering many old things such as money, small statues, and such.

The Babylonians have many artifacts adorned with the face of King Hammurabi. There are local merchants with booths setup just outside the inner perimeter wall of the FOB but inside the outer wall. I had some time to kill while the colonel was flying so I visited the booths and began inquiring about old coins from ancient Babylon. It took me quite a while but I finally found a gentleman that knew what I was searching for. He pulled a small pouch from his pocket and opened it. Inside he had approximately ten coins that appeared very old and worn. I purchased a coin from him that is said to be over a thousand years old. It looks authentic to me but I know little about coins. There were some young kids there that seemed so happy in spite of all the violence around them. "Mista Mista!" one boy shouted, "Me carry?" He carried my bag and followed me around for a minute or so. I gave him a dollar and he went on his way. I couldn't resist taking their pictures and they loved the attention. It's sad that they have to grow up in such a harsh environment.

Above - The young Iraqi boy that carried my bag

Above - Babylonian ruins – You can see the rebuilt city walls in the distance.

Inside the walls

Above – Hieroglyphics in the walls

The rebuilt city is now only a tourist attraction.
The cost for the tour was one dollar

7 May 2004

With no obligations today, I went for breakfast at the chow hall down on the Tigris River. They have patio seating just outside the building beside the water. It's very nice in the morning as the sun is rising and the heat hasn't yet scorched the air. Palm trees are plentiful and wave in the breeze. I later learned that Saddam had executed an entire soccer team on this same patio for losing a game while representing Iraq. As the story goes, he lined them up along the water's edge, executed each of them, and kicked their bodies into the river to float away.

I walked back to the Tactical Operations Center (TOC) to catch up on the latest intel and found out that we had killed forty-two more of the Sadr militia. Chalk one up for the good guys. This apparently pissed them off so they mortared one of the FOBs in the city of Ah-Najaf and we retaliated by raining them with 155mm mortars followed by a 500lb bomb.

After visiting with one of the pilots for awhile, he and I went back to the public market to buy a pistol holster. Mine had broken while in Kuwait and I'm currently using an old Vietnam era holster that was somehow still in the troop supplies. I paid $11 for the new one. It was nice to finally have a comfortable holster again. I remember seeing a guy at the market that was very disabled. His left hand was missing and he had necklaces hanging from the stub along his forearm. His face had been burned in the past and he was difficult to look at. Although I didn't know his story, I was sympathetic for him. There are many different people here, each with a story. Now I have a story. I never thought I'd be standing in the middle of Babylon. After spending my day soaking up the atmosphere, I watched a movie with some of the Nomad pilots, and then turned in for the night.

My room in Babylon

8 May 2004

I woke up this morning to the sound of rockets flying over our FOB. Someone launched a rocket attack against a camp near Babylon. I could hear the rockets whistling overhead and the impact as they came down. Having heard this many times before, I knew they weren't close so I rolled over and went back to sleep. I eventually got up at 0700 and shaved before going to breakfast. After breakfast, our flight prepared for our departure back to Al-Kut. With a large hangar in Al-Kut, our FOB is much better equipped for major maintenance so we're flying a broken aircraft back to our FOB for repair. The bird was missing an electrical cable for one of the weapon systems and the ground to air radar jammer was inoperative. Not the best feeling flying a broken ship but it had to be done. We made it back to our FOB in the crippled aircraft without issue. After returning, I stowed my flight gear and rested a bit.

The pilot I'm normally crewed with flew with another pilot while I was gone and their aircraft had a hydraulic system failure during their mission. In the OH58D, that's similar to driving an old car without power steering except you're trying to fly with a great deal of precision. They were 30 miles from the airfield when it occurred so it wasn't an easy flight back. Both were very exhausted but managed to land the aircraft without any damage. With this type of failure, the helicopter is landed like that of an airplane, with some forward airspeed. The landing gear has metal strips on each skid tube called skid shoes. These shoes protect the landing gear from damage in a situation like this when a run-on landing is necessary. Two of the hydraulic lines had rubbed together and chaffed one to the point of failure. The hydraulic system has about 1000 psi of pressure in the lines so even a pin hole can lead to a large rupture. Close calls seem to be the norm now so I thank God daily for his protection. I think I'll rest for a while, catch a shower, and hit the bunk for the night.

9 May 2004

Today is Mothers Day and I miss being able to be home to wish my mom a Happy Mother's Day. I miss giving my wife the same courtesy through our kids. I tried checking my email but the system is not working well. I wanted to send more email but I guess that will wait until later. I'm scheduled to fly this afternoon and it's really hot out today. I've never appreciated cloud cover like I have in Iraq. Even just a few minutes of shade can cool the air ten degrees. Ten degrees doesn't seem like much until you're wearing a full ensemble of flight gear covering every square inch of your body and three layers on your core body. Nomex flight suits are hot enough as it is, but we also wear a layer of body armor with a ceramic plate over the nomex and a survival vest over the body armor. The survival vest is equipped with items critical for survival and evasion in a downed aircraft scenario.

I flew my mission and all went well. After returning, I was told I would fly an early morning mission so I ran to the command center to attend the air mission briefing for that and debrief the intelligence officer on our previous flight. Shortly afterward, I was taken off the mission to allow someone who hasn't flown in a while get back in the saddle. The rest is nice but I'd rather have the flight experience.

Tomorrow I should have time to pickup my laundry, maybe go for a run, and do some reading. I have a lot of stuff that I want to read and believe it or not, I don't find the time. Many of the simple things in life back home take a long time here. It takes an hour or two to eat chow or take a shower because of the distance they are from our living area. It's a twenty minute brisk walk to the chow hall just over a mile away. The time passes quickly. I need to start studying for my annual flight evaluation. It's due shortly after we return home. I was able to email home. I hope my mother and wife have a great Mother's Day. I played a game of cards with the boys after checking my email and showered around 2200 hours. Now it's time to turn in.

10 May 2004

Mother's Day has passed, which is another day gone and a day closer to seeing my girls. After being pulled off the morning mission, I slept in and enjoyed it. I finished a movie and then looked at some pictures from home while listening to music. Our new firefighters just arrived on the FOB and I was tasked to help train them on aircraft incidents and how to perform an emergency shutdown on our aircraft, if the need should arise. The training went well. Most of them had never received training on the removal of injured personnel from an aircraft so we walked through it, step by step.

We ordered some large white circus style tents to house some of our troops and those were finished today. The pilots all continued living in the hangar bays while the crew chiefs moved into the tents. This freed up a lot of space for the pilots. With the additional space, it allowed the pilots to split up in separate rooms based on shift. Day shift used one bay while night shift used another. Until now, we've all been mixed, making it very difficult to get quality sleep. It should all work out as long as they don't add more people to our bays.

Tomorrow I intend to get up early and go to breakfast with the Lieutenant. I don't fly tomorrow so I'll be building a couple things for my living area and studying some aviation material. I was going to email home but the internet has not been working tonight. We get the satellite phone at midnight so hopefully I can make a call tomorrow. I need to talk with my girls and I'd like to get Bertie to send me some pop-tarts and granola bars.

We are making steady progress with the war effort. Some recent missions played out well with no friendly casualties. I hope and pray that future missions continue the same.

11 May 2004

I was up by 0500 this morning and thought I'd go for a hot breakfast. I went to check my email and found the system is still inoperative. The heat this morning was much higher than normal. The clouds came in late yesterday, causing a blanket of insulation to keep the heat close to the ground and wow did it ever. It was very warm this morning already. It's about 95F this morning and the sun hasn't even risen.

The command center has national TV service setup now in the common area. The big thing in the news right now is abuse of Iraqi prisoners. The media is tearing this one up. It's likely that it's being over-emphasized as usual. I don't feel sorry for them. Many aren't Iraqi but have come to Iraq to terrorize either Iraqis or Americans. From what I understand, many are from Saudi Arabia.

I worked on designing a digital kneeboard mission sheet template for all the air missions. I made some changes to the format so it could be printed on card stock paper. It has a more professional look and the thicker paper will endure the elements much better. Afterward, I helped the LT build a shelf for his living space. If you want furniture here, you have to build it yourself.

We ate chow around 1730 then had a troop meeting. Most of my day was spent trying to improve our internet connection. I called home today and talked to Bertie. She's doing well. Her sister just had a baby so that helps keep her mind off of me. The little guy looks so cute. His name is Ethan and she's of course very proud of him. I also called my mom. She was at work but was able to talk for a few minutes. Mom was glad to hear from me and me from her. Dad did not answer the home phone so he was probably at his office working. It was nice talking to my mom. Hopefully hearing from me cheered her up.

12 May 2004

I slept until 0930 and then watched a movie in bed until 1100 hours. I lazily got up went to chow. After lunch chow, I came back to the hangar and made a laundry drop since they would be closed by the time my mission was complete.

Our air mission went well. We found more unexpended ordnance lying in an old Iraqi battle position for a tank, just waiting for someone to pick it up. Most of it was 155mm rounds and some mortars. There is so much of this stuff just lying around. Our ground squadron also found an improvised explosive device (IED) today, thanks to an informant. It was several 155mm rounds daisy chained together, exactly like the ones we found. They are deadly to our ground forces when in the wrong hands. We did well today and probably saved several lives. When we find these things, we call the Tactical Operations Center (TOC) and report it with an exact location so our EOD team can retrieve them and either destroy them in place or move them to the FOB and destroy them at a later time.

Many locals were waving at us today. We dropped a soccer ball and some crayons to the less fortunate kids living in tents. They loved it. I enjoy seeing the kids smile. I think I'll ask Bertie to send me some stuff to give to the kids. I did very little other than fly today. We flew just over five hours and I'm exhausted from the heat.

The Iraqis love flying kites and once airborne, they tie them off on their houses and they'll sometimes fly for days, depending on the weather. Some extend as high as 400 feet. I often wonder if it's intended to keep us from flying over their houses. We typically fly at an altitude of 50' to 300' and at airspeeds of approximately 100 mph. We often find kite string in the rotor system during our post-flight inspections.

13 May 2004

I've been back in Iraq one month today. It's gone by rather quickly. We've flown quite a few missions, however I've logged fewer flight hours than I did the five weeks I was flying in Baghdad. With my experience building rapidly, it shouldn't take me long to attain my pilot in command (PIC) rating. With the mission complexity, the pilot in command sign-off isn't handed out easily. I don't expect to see that happen here in Iraq but it shouldn't be long after returning home.

I am beginning my study sessions today. I made a schedule that will allow me to cover every topic in detail before my check-ride. I was supposed to fly with one of our new PICs today. He was anxious to fly the mission but bad weather kept us on the ground. The dust storms stir the dirt up tremendously. That's typical spring weather in Iraq.

I checked my email and sent one home this morning. Yeah, the internet is working today! Many times it doesn't work at all. I need to email my brother; he's probably wondering how I'm adapting. We went around the FOB today trying to track down a guy that knows something about getting a better internet connection setup. We finally found him but he has a guy that works for him who does all the manual labor of setting it up and he was gone. Nothing is easy here. During our trek, I found a barber shop and stopped for a haircut. I had it cut very short just before leaving Kuwait so I would be set for a while. It has a month on it now so feels pretty shaggy. I'm flying early tomorrow so I should turn in for the night.

14 May 2004

I was up at 0415 this morning and boy that felt early. Wiping my eyes, I grabbed some coffee, went to the command center for the mission brief, and then off to chow before flying. I was scheduled to fly four hours today and ended up with almost five. Some of the explosives we found and reported had been picked up by someone other than our EOD team, so we spent a little time investigating the area in attempt to find something that would lead us to who may have taken them. We flew over many people out by an old military ammo cache and they all seemed harmless. They were spending their time disassembling a building brick by brick. They're quite the scavengers. To be caught picking up any ordnance lying on the ground could be a death trap for them, according to our rules of engagement.

We continued our mission providing air support for a convoy moving on a main service route (MSR). As we flew down one route, we saw that the Iraqi Police Force (IPF) was up ahead securing an area around an IED on the road. We guided the convoy around the area to avoid it and continued. Our Ukrainian coalition partners arrived and took control of the area from the IPF and used their EOD team to defuse the roadside bomb. It was a great mission today. The enemy plan was foiled and no one was hurt. Chalk one up for the good guys.

When we returned, I found out that I had been scheduled to fly tomorrow's mission with the staff pilots. Staff pilots are the command officers that are also pilots but fly little compared to Warrant Officers, in most cases. Staff aviators' primary role is to lead the squadron or troop. I'll be flying with Captain Coyle this time. He seems alright and hopefully easy to fly with.

I received a package from home today! It was on my bunk when I came in from flying and was filled with all kinds of goodies to boost my moral. It had pictures, a handkerchief from Roberta with her perfume on it, some ice tea and sugar, and a letter from the kids. It was the highlight of my day. I wanted to email her right away but as usual, the internet wasn't working. It doesn't work well during the day when it's hot. I'll try again tonight. After my haircut yesterday, I feel much better. I had one of the locals cut it which is a little scary. I'm always paranoid that one of them may try to kill me with the scissors or razor. They also have this strange technique of plucking eyebrows with a twisted elastic string. It looks painful so I passed on the additional beautification and headed back to the hangar. I'm not sure if it makes a difference, but I only go to the barber in my PT (physical training) clothes so they couldn't tell I was a pilot. Pilots are high valued individuals to the enemy.

15 May 2004

I got up early this morning to fly a recon mission but the weather had other plans. The wind was strong and blew up a large dust storm. Dust storms are bad here. The wind has been blowing up to 50 mph. I was out on the flight line last night loading some navigation data into the aircraft's on board computer for today's mission and saw an explosion across the airfield. It startled me but it was only the EOD team. They destroyed the live rounds we found the day prior. We are finding so many rounds on each mission that it's difficult for

EOD to collect and destroy them all. They are being picked up by Sadr's militia, the Mahdi Army, and they're using them to build large bombs. Many IEDs have artillery rounds daisy chained together for multiple explosions or one large one. These bombs and IEDs have killed many soldiers and civilians.

Another local city saw much fighting last night and three U.S. soldiers were killed. The gunships rain fire down on the face of the enemy so they try to retaliate. The soldiers that died were from First Squadron Second ACR. Our Regiment is killing many of their militia group but the Mahdi Army is very fluid in numbers. They won't win the fight but the question is, how many will die before we break them?

On a good note, I was able to email Bertie today and she said the rainy season has started back home. She said it only quits long enough for the water to subside and then it rains again. I hung the pictures she sent me from home and many people commented on the pictures of my kids. I have a great family that I'm thankful for and proud of.

I guess I'll do a laundry swap today. It's nice to have a laundry service again. We received a load of lumber today. The hangar looks like a furniture manufacturing plant. I love the smell of fresh lumber. We immediately went to work building things. Small pieces of furniture like tables, shelves, and bookcases make life nicer for us.

I've been thinking about the flag I've been flying for my parents. I've taken that flag on many missions and it will go on several more. I know it's only something very small but it means the world to them.

16 May 2004

Our mission was cancelled today due to administrative reasons so I didn't fly again today. I was rescheduled for tomorrow. I used the time to study some aviation related material in preparation for my annual check-ride. I heard a new shop opened up on the FOB so we visited, played some ping pong, and tried to get on the internet but the line was too long. I wasn't patient enough to wait so I picked up laundry and went back to the hangar. I sent a couple emails and made up some of the tea the girls sent me before hitting the cot for some rest. There's nothing like some southern sweet tea for a little pick-me-up. Ahhhh!

17 May 2004

At 0230, I woke up to the sound of mortar fire. The explosions shook the building! After several rounds impacted, I jumped up and went down to the TOC to find out what was going on. The XO said it was the Ukrainians firing for some reason. My heart was pounding out of my chest. I thought our FOB was under attack. A sheet metal building is about the worst type building to be in during an attack. The metal makes it sound much worse than it is. After hearing that it was friendly fire, I went back to bed and slept until about 0930.

I had a cup of coffee and then went for a shower. After cleaning up, I sent an email to Bertie and watched the news as they spoke of the car bomb that had just exploded in the green zone of Baghdad. They didn't give a casualty count. It's sad that there are so many that are Saddam loyalists and are resisting democracy. I spent the remainder of the day working on a computerized document template for the Crew Chiefs and flew from evening into the night. I was very tired after returning. I also found out that I'll be flying with the Colonel again tomorrow. It seems that they always pair me up with the Colonel when he decides to fly a mission with us. I get the impression that none of the other pilots want to fly with him. Since I have little seniority, I always get the short straw. Oh well, I'll fly with him. We get along great and I get lots of good inside information in the process, like when we may be going home. It looks like we'll be here another month but not more than two.

18 May 2004

I'm scheduled for a morning mission today so I got up early enough to walk down to the chow tent for a hot breakfast. It wasn't bad. I made the long walk back to the TOC for the air mission brief. The Colonel gave me seat choice so I picked the right seat since it's primarily flying the aircraft. He usually wants to fly right but I jumped at the chance to chauffeur him around the AO. Our mission included a sneak and peek up on the Iranian border. A small border town on the Iraq side was reported to have an anti-aircraft gun that we wanted to verify was still decommissioned.

After a quick trip down south to recon a supply route, we made a turn in the FARP to top the birds off with fuel. Our team then flew northeast of Al-Kut and travelled a route that took us about twenty kilometers south of an Iranian border town. We departed the route and picked up a river bed to maintain the element of surprise upon arrival. The river was fairly wide and allowed us stay below the ground level but above the water. We followed the river at nap of the earth (NOE) altitudes for the last twenty kilometers. Snaking our way closer to the border, our senses were dialled in and our focus was sharp. We saw one man in the river and as we flew over, he dove under water. I'm not sure if he thought we would hit him or shoot him but we were at about ten to fifteen feet above the water at that time.

Upon arrival of the border town, we rapidly climbed from the river bed. Flying from out of nowhere at nearly 100 mph, I'm sure we surprised the whole town. It's a quiet few seconds during that nose high climb. You only hear the whine of the turbine and a few pops from the blades as they cut through the air. Once at altitude, I banked the aircraft over so the Colonel could get a look at the anti-aircraft gun and maintained an aerial position that would allow me to quickly mask the aircraft if they began firing at us. Fortunately, the anti-aircraft weapon we were there to recon was still out of commission and didn't appear to have been tampered with. We took a good look at the town and found nothing out of the ordinary. We always felt that there was smuggling activity from Iran in this area, but didn't have the man power on the ground to put anyone up here. The flight time enroute was approximately 45 minutes.

Just to our east was the Iranian border. The mountains in the distance are lined with air defense positions that are well lit at night. Sometimes the Iranians would transmit to us on our emergency frequency and in a foreign accent say "You are approaching Iranian airspace." We are well versed on the border location and were never concerned that we would inadvertently breach their territory. We conducted a route recon on the way back to Al-Kut but found nothing to report. Our flight was five and a half hours and my legs were feeling it.

I'm beginning to have issues with my left leg after three to four hours of flying. It's going to sleep or something. I think my circulation gets stagnant after four hours or so. The seats of the Kiowa are made of a mesh material that is stretched over a metal frame. There is no padding so I'm using an old seat pad from a Blackhawk which helps but doesn't completely stop the pain or numbness. It does give me another hour of flight time before the issues start. After returning, I took a nap and then did a laundry drop before dinner chow.

Entry point of the river – N.O.E. to the border town
Background - Mountains deep into Syria

On the way back from chow, one of the Polish guys stopped me and Captain Chapman and asked us if we were interested in trading American goods for something they have. We were interested so we showed him a few items we could part with and hopefully today, we'll be able to make a trade. We arranged to meet him at 1800 at an intersection just outside the hangar. We went to our daily troop update brief then I went for a shower. After my shower, I met with the Polish soldier and we made a trade. I had an extra pair of Wiley-X sunglasses that I traded him for a military issue leather map case. The Polish soldiers have been very kind and are proud to be part of the effort in Iraq. Their primary function on the FOB is air traffic control (ATC). There is a major language barrier when talking to them so we have a very keen awareness at all times. Several times when inbound to land at the FOB, I'll call two mile final for the runway and they'll respond with "You are cleared for takeoff." Umm....This doesn't give me a warm fuzzy feeling. I immediately looked for departing traffic only to realize that he's talking to me! Our situational awareness has to be dialled in all the time, mostly at night when C-130s are landing blacked out.

Above – Polish map case and picture of the inside cover flap

19 May 2004

I woke up about 0730 for an outhouse run and then went ahead and dressed for the day. We finally have more port-a-johns and a shower building by our hangar. The shower facility is unisex and has time blocks for males and females. I'm currently waiting for the women to finish up so I can go shave. With only a handful of port-a-johns up until now, they were full within a day or so and became very unpleasant to be in. The challenge with a one piece flight suite was keeping it off the nasty urine covered floor until you finished.

After this cup of coffee is gone, I'll see if I can send some email home. I'm on standby for a mission today so I have to be available but can still get some stuff done. I picked up a batch of laundry and was able to mail a CD home with my pictures on it. Bertie will appreciate getting the photos. Hopefully none of them scare her. We had an increased threat of chemical warfare today so now we have to carry our gas masks around and have it accessible within ten minutes. I can't imagine being caught in a chemical biological attack, mostly if I wasn't properly protected. What a terrible way to die. This has been another method used by Saddam to terrorize people.

20 May 2004

The month of May is two-thirds gone. I hope the time continues to pass quickly. I'm on standby again today, but our schedule should be back to normal tomorrow. I've been trying to check my email but the system is not working. Gas prices are still climbing back home. Dad says the prediction is to see $3 per gallon before the end of summer. I hope that doesn't happen. God has taken care of me and my family while I've been here. We are making a little more money while I'm deployed. It doesn't make it okay, but it does add a positive light to the situation.

Bertie is doing a fantastic job at home and I'm proud of her for her strength and ability to step up to the challenge. I knew she had it in her; she just needed the chance to do it. Most military wives get that chance before their time is up. It is definitely making us stronger. We've never been separated this long. So far, we've had no arguments while I've been away. I try to keep my mind off of my girls because it just makes me miss home even more. I have to spend some time, at least once a week, looking at pictures and I find myself re-reading the letters they send me.

21 May 2004

Well, I'm on standby again! This will be my third day without flying. I'll kill some time working on the Aviation Mission Planning (AMPS) computer today. It's been broken for a while now. I emailed the support team so now I'm waiting for their response. That should keep me busy much of the day.

I made another laundry drop today and tonight out LT assigned me the duty of completing the monthly sensitive items inventory. It sucks being the new guy. A sensitive items inventory is nearly self-explanatory but I'll expound. It includes visually accounting for all weapon systems on each aircraft, all personal weapons (M-9, M-4, & M-16), night vision goggles, and the classified items in each aircraft. Each is tracked by serial number so I'll have to walk around and track down serial numbers for everything on the list. This task will take several days. Now I know why people loathe doing it. I need to get some rest, I'm flying early tomorrow.

22 May 2004

We had a 0530 crank time so I got up at 0400 to make sure I had time for enough coffee to prop my eyes open. We had some bad weather moving in but we still managed to complete our mission of reconning a few routes. During the mission, I noticed some pressure on the cyclic that felt abnormal. I wondered if it was a wire binding it up because when we landed, we had some electrical problems causing all types of abnormal indications in the cockpit. We made it back safely, thank the Lord.

The wind was beginning to pick up and the storm that was forecast was finally blowing in. We began seeing lightning near the FOB so they closed the fueling point. We de-armed the aircraft, parked, and shutdown just before the rain hit. We managed to fly nearly four hours and completed our mission just in time.

Our FOB hosted a comedy show tonight and it was a great time. The last comic was absolutely hilarious. He made light of the different roles of military leadership and had me in tears. It was a great morale booster for all the troops. I think the SGM was a little perturbed with his jokes since he found himself at the punchline of many of them. Oh well, he'll get over it. He made fun of all ranks - Captains, Warrants and Enlisted alike. He even poked a little fun at the pilots saying we walk like we have a stick up our ass and thought we were better than everyone. A few of the upper enlisted guys were offended but most loved it. His depiction of E4s driving the short bus to Lieutenants walking around aimlessly gave us all something to laugh at. It was a nice stress reliever after having our tour extended.

23 May 2004

It's Sunday and I'm not scheduled to fly. They downsized our mission load a bit on Sunday. I started the morning with a cup of coffee as usual and tried to check my email but the internet wasn't working well enough. We ate lunch around 1200 hours. The food has been good. Recently, a new contractor took over the chow hall and we're much happier.

In the afternoon, I continued working on the sensitive items inventory before going to dinner chow. After filling my belly, I spent some time smoking my pipe - one of the prize commodities Bertie sent in my care package. This pipe was given to me by Jason, my closest friend back home. I smoked my pipe and recapped the two hundred combat hours I had accumulated.

After a relaxing smoke, I did some engine run-ups for our maintenance test pilot and then reviewed some technical information about the aircraft. It's good to stay sharp and keep each other on our game, mostly when you have a maintenance pilot at your disposal to answer questions.

I ran into one of my Basic Training buddies at the chow hall and invited him to visit the flight line. In the evening, he stopped by to see me. He is an EOD soldier and on a team that we've kept extremely busy by finding rogue ordnance lying around. He was well versed in his field and I was interested in his work so we messed with some C4 and all the cool explosive gadgets they carry. They have a robot and a .50 caliber sniper rifle. Cool toys for cool jobs. Most only see this stuff on TV. Wow, what an experience! He was more interested in the aviation side of the war since he's seen so much on the ground already. Many will never see Iraq from the air or faces of the people across this country. There's an array of expressions and mixed emotions from the Iraqis. Some are excited, others are terrified. Many times when we fly over, they will run as if they think we'll kill them for no reason and others wave with overwhelming joy. We drop candy and toys to the kids. The poor people that live in tents fair the best with my heart. I enjoy giving them a brighter day. Maybe it temporarily fills the void from not seeing my kids, it makes me smile none the less.

24 May 2004

I was on QRF today so it was a good opportunity for me to complete much of the inventory, which I did before lunch. Our maintenance pilot needed someone to complete a test flight with him so I jumped at the chance to help out and snatch some flight time of course. He was originally taking one of the crew chiefs but after hearing that a C-130 was engaged with a surface to air missile last night, he quickly changed his mind and decided he would stay on the ground. We took off and flew an hour, testing the mast mounted sight. It was uneventful but the sight was definitely not working properly. After landing, I went to dinner chow before making my way back to the TOC for the Troop Update Brief. The briefing held no surprises and was filled with run-of-the-mill information. The FOB is slowly developing into a liveable area. Showers, hot meals, and air conditioning make the days much easier to bear. It took us a week to get air conditioning in our sleep rooms after arrival. The sweat would run off of me and attract mosquitoes. After the first night, I used a heavily concentrated deet mixture which seemed to stop the bites, but nothing stopped the continual buzzing around my head. Most of the soldiers were issued mosquito nets before being deployed. By the time I reached the unit, Central Issue Facility (CIF) was out of stock. I think I was the only person without a mosquito net over my cot.

Our living area before air conditioning

25 May 2004

I was scheduled for an early morning mission so I got up about 0500, just in time for some coffee. We received our mission brief at the TOC and then moved to the flight line for departure. During our mission, we had a video card failure in the aircraft's on-board computer, causing my display to black out. We landed on an abandon airfield about forty miles south of the FOB to troubleshoot it. It was definitely dead and there was no bringing it back online. We flew back to the FOB, swapped aircraft, and finished the mission. During the last portion of our flight, the trail aircraft hit a flock of birds that flew up through the rotor system and the aircraft began vibrating horribly. A flock of pigeons had hit every one of the main rotor blades and damaged them all. The damage threw the rotor system out of balance so the pilot slowed the aircraft to a safe speed and we returned to the airfield where he landed without further issue. Fortunately, we were only five miles from the FOB when it occurred.

I grabbed some lunch chow before taking a short nap and flying a test flight with our maintenance pilot. Not a bad day overall. I flew seven hours today. What a great day for flying. After dinner chow, I checked my email and took a much needed shower. My clothes were soiled with white salt lines from the continuous sweat that poured because of the desert sun. After cleaning up, I knocked out more of the sensitive items inventory before hitting the sack. I'll be glad when I get this inventory completed.

Chapter 9

Saddam's Revenge

26 May 2004

Today has been the hottest since I've been here. It was 110 degrees. We had a four hour mission to fly in mid-afternoon and I was drained after the first hour. We talked about how it should cool down as the day progressed but there was no relief. I went through all my water within the first two hours of the mission. We conducted a route recon northeast of the FOB up to the Iranian border and took some great pictures of the mountains in Iran. Surprisingly, we also spotted some flamingos in the sunset. This is one place I never expected to see flamingos. Maybe I was hallucinating.

One of our mission tasks was to recon two bridges that had been blown up and get some photos. We programmed the grids into the aircraft and upon arrival confirmed the information from our intelligence group. Both bridges had sections that were destroyed. Many of the bridges are guarded by local police and aren't the most reliable with security measures. This continues to be a problem throughout Iraq.

We recently had a soldier on our FOB that was en route to his FOB and was delayed for a short time, waiting to be air lifted. He had been hit by mortar shrapnel and sent out of Iraq for treatment. He had recovered enough to return to the fight and was on his way back to his unit. I asked him to share his experience and he did. He started with the first mortar of the set.

"When the first one hit the FOB, I began to move toward the bunker. After my first step, a mortar landed just beside me and blew up. I tried to

run for the bunker at that point and fell down. I really didn't think much of it at first. I was shell shocked, blinded, and deaf for a minute. I could go no further. I looked down and saw my legs covered in blood. I didn't feel much pain initially. I was taken to the medics and after many x-rays, they found that I now have enough metal in my legs to equal the size of half a baseball."

It had been 5 days since the attack when I saw his legs. There were hundreds of tiny scabs on his legs that ran from his ankles to just short of his groin. He was thankful that his jewels were still intact. I'm sure this will be painful for years to come. You may ask the question, why didn't he just run after the first round hit? The answer is simple; you get so used to them coming in that it's a painful routine to always go to the bunker. Many times, I lay in bed at night during attacks and if the rounds weren't close to me, I wouldn't even get off my cot. It feels like roulette.

27 May 2004

The inventory is almost complete. I have a few things left but the majority is done. What a pain. I hope I never have to do that again. I went to supper and something has been hitting my stomach hard. I have had cramps and the diarrhea for three days. I woke up with a headache this morning too. I managed to stay up and go to midnight chow after an hour long workout. Even the midnight chow got to me. I was talking on the phone to dad and it hit me hard. I had to go right then. I ran to the port-a-john and barely made it in time. I went back and called him again. He said my grandma will not be driving anymore. I should give her a call and see how she is. She is nearly 90 now. I made another trip to the port-a-john. I think I've caught something.

28-31 May 2004

I visited the flight doc and was diagnosed with having a bacterial infection, a.k.a. Saddam's revenge. Doc gave me three Cipro. I'm trying to reset my sleep to a night schedule. My flights will be during a much cooler period. Since I'm sick anyway, I'll use this time to reset my schedule.

I was able to meet Bertie on chat while she was at her parents' house. I was happy to talk with her and the girls. It sounds like they're really enjoying themselves in Missouri. She is in love with her sister's new baby and has hinted around about having another baby. I'm not keen on the idea. Two is the perfect number in my opinion.

We had a bunch of new equipment issued to us on the 30th. It was a whole bag of goodies: new boots, sand and dust glasses, gloves, and fleece coats. Yes you heard me right, fleece in the summer. Why, because it's the Army and who knows what goes on in their mind. It was like the Christmas day scene of The Christmas Story. Soldiers love new gadgets. They laugh at $1000 extra per month but give them new gadgets and they'll stay in Iraq another year. It was a morale booster to say the least. We were skeptical that the promise to return home by July was true. Why else would they be issuing us winter gear?

1 June 2004

We're now into June so we can actually say we are going home next month. The time is winding down and people are beginning to talk about it daily. I hear both extremes from "We'll be packing up by end of June" to "I'm not planning on being home until October." All things indicate that we'll be back by end of July so I won't miss Madison's birthday. I'm excited to see the family.

I'm on QRF tonight so I'll most likely be sitting around doing very little. I'll use the time to do blind cockpit drills. Blind cockpit drills consist of sitting in the aircraft in the dark and working through emergency procedures with no lights. The key is to be able to identify key circuit breakers and switches without using lighting or turning to look at them. This process allows a pilot to quickly work through an emergency situation without delay. Additionally, I'll also refresh my memorization on aircraft limitations and memorize every step for any given emergency.

It's 0300 and I'm finding it difficult to stay awake. It's even worse while trying to study. I'll hang in there until 0500 and stay up for breakfast. Another night or two and I'll be acclimated to the night schedule.

2 June 2004

I woke up about 1230. I need a little more sleep but that's not gonna happen. I'll mess with my computer for a while, maybe that will put me to sleep. Wow! I just received my other package from home, so now I'm really awake. What a great morale booster! My girls picked out some goodies for me. Everyone enjoys getting packages from home. Words can't express how simple things like tea, sugar, and sweets can make our days go by so much better.

I'm on the flight schedule again tonight. I'm still racking up the hours. My goal is to have more combat hours than training hours before I return home. At the current pace I should have no problem hitting that goal.

Each evening, we have an update brief for all the things that happened in the past 24 hours, most of which is usually bad news about attacks, injuries, and deaths. We undoubtedly always kill more of them than they kill of us but we don't want to lose even one more soldier. Sometimes I'd rather not hear the intelligence summaries.

They try to kill us in many different ways and that's probably traumatic to family and friends back home but here that's our life from one day to the next. A casual conversations before a flight may be, "I heard there was another surface to air missile (SAM) engagement so watch yourself around that area" with a response of "I always do, besides they're not a good shot anyway." It's been a big adjustment but now, it's just a way of life, literally. I try not to compare it to back home but can't help it at times. I'm reading a book that my wife sent me about one's purpose in life and it's pretty good. It eases the stress some because I realize that I'm here for a specific purpose and I won't be gone until God says it's time. That alone gives me a lot to look forward to. I'm reminded of a scripture that says "if God be for me, then who can be against me". It renews my confidence daily.

3 June 2004

While I was flying a mission from the evening to early morning, I thought of how close we are to leaving this place. It reminded me of home and gave me a burst of energy. The day crews are firing missiles for practice today. I'd like to participate but I'd rather stay on night shift and out of the scorching temps. I'll have a chance to fire another day. We are watching the TV series "24" after flying each day. We are about half way thru the series. It's popular in the troop and helps pass the time. Time for a shower then I'll sleep away the hottest part of the day.

4 June 2004

I didn't sleep well today. I still haven't quite adjusted to my night schedule. I'm fine during missions but I drop off like a dead alkaline battery after returning. My body clock is screwed up and I'm flying again tonight. Right now, we have teams flying 12 of the 24 hours in the day, if not more. I expect we'll be getting ready for redeployment home soon, hopefully for good this time. I sifted through a bunch of photos from home today, reminiscing of my family time back home. I pray that God brings us home safely.

I'm flying right seat tonight, which takes a little more mental preparation for me since I'm a junior pilot. The mission is more taxing on right seat pilot at night because he's on the controls nearly the entire four and a half hour flight so it's imperative to have the keenest of senses and maintain mental focus. We fly low and fast so adding the environmental factor of darkness and heat into the mix is extremely dangerous since their towers and obstacles are unlit. We do have the advantage of the night vision goggles but even the NVGs have limitations. The day flights are different in that both pilots are fighting for the stick. Flying low and fast limits the visual exposure to our enemy. Our noise signature is masked by the low altitude and we can maintain the element of surprise. Time to go burn some jet fuel!

5 June 2004

We finished our mission and all went well. We saw a couple tracers come up in our vicinity but that's about it. I'm sure they were taking pot shots at us. Lucky for us, they can't see us at night. We flew up to the Iranian border and checked it out. It's interesting to see at night. Their security positions are well lit along the top of the mountain ridge, just inside the border. We finished our route recon up north along the border and took the river to return south. Upon reaching Al-Kut, the TOC radioed that there were mortars being fired and not to approach the FOB. They didn't want us to approach the airfield until they verified what was going on. The Ukrainians were firing some mortar rounds from the FOB to register their equipment. The Ukrainians are responsible for our perimeter security on the FOB and they test their weapons from time to time. They never felt it necessary to let us know ahead of time. The impact of a mortar round near the FOB shakes the hangar like thunder from a bolt of lightning ten feet away. I recall the Ukrainians shooting at some men trying to sneak on the FOB one night. The hangar shook and the sheet metal rattled. My heart beat a little harder at the impact of each mortar. I didn't get up until about the tenth mortar round hit, then I thought I'd better go see what was going on. There was an insurgent trying to sneak some explosives onto the FOB.

The news speaks of peace talks in Najaf lately. They're under a cease fire agreement. Our ground squadrons have consistently killed Mahdi militia members in Najaf. The militia is slowly migrating into our area of operation and seem to build in numbers anywhere we have soldiers. Our ground troops are well experienced with the warfare and tactics of Sadr's army, so it's second nature to them at this point and presents little challenge. It's still an adrenalin rush when things flare up. Who wouldn't get worked up a little with someone trying to kill you?

I called Roberta at 0200 my time but there was no answer. She took the girls from Missouri and drove over to Ft Campbell Kentucky to visit some friends that are stationed there. That's a big deal for her; she normally doesn't venture too far from home. I hope she made it ok. In spite of all the negative effects of the deployment, it's been a growth opportunity for her.

6 June 2004

I didn't sleep well again today. I only slept until 1230. I'm not flying tonight so it won't be a safety issue. I did some reading and dozed here and there. I've heard rumors that one of the pilots who joined our unit late in the deployment may be sent to another unit to back-fill an open pilot slot and won't return home on schedule. I suspect that's just a vicious rumor and isn't likely to happen. However, it has me concerned because I joined 4/2 ACR around the same time frame. It wouldn't make sense because our unit is losing several pilots already and that will leave them with a skeleton crew back in the States. It gave me a bad feeling for several hours today. The thought of being extended even longer is like a poison; it makes you burn inside. I can't stand laying disappointment on my family like that. I'm dedicated to serving my country but when you mess with people's families and lives like that, it kills motivation. I had a headache after hearing that. I don't know if it was that or something else but I took a Motrin and laid down in hopes of forgetting it and falling asleep. Some people start rumors simply because it's a game to them.

I woke up for breakfast about 0500. I'll try to get a haircut after 0800 when the barber shop opens. Then I'll shower and sleep, hopefully. I tried getting on the internet this morning but it was down as usual. I tried calling Bertie last night but she didn't have her phone on. I miss talking to her. I enjoy listening to her voice.

Well, much has happened since my last writing. The rumor wasn't true and was someone's game. We received news that we are leaving the last week of June which is a couple of weeks earlier than expected, praise God. It's time for some chow.

I checked my email to see if Bertie had sent any mail back but she hasn't. The last few days have been such an emotional roller coaster. Bad news puts me at the bottom of the lowest valley and good news lifts me beyond the highest peak. It reminds me of an old Baptist hymn I sang as a child "It reaches to the highest mountain and flows to the lowest valley, the blood that gives me strength from day to day it will never lose its power". This experience has changed my life and my outlook on life. I am a better person for it. God is great all the time.

I've been reading a new book about the tests that God puts us thru to judge our reactions. These tests are for preparation for our eternal lives with him. It's interesting and I'm learning how to determine my purpose in life. *I realize that this life is just an interview for eternity.*

7 June 2004

Our mission ran late tonight. We were only supposed to fly one bag of gas and it turned into two. I was tired afterward. The moon illumination was extremely low early in the flight so it was hard to see, even with the night vision goggles. We provided aerial security for a 140 truck convoy of Iraqi Civil Defense Corps (ICDC) personnel. That's a long line of slow moving headlights from the sky. Convoys this size will typically travel at night to reduce the risk of an attack. We have the advantage at night and the capability of night vision, which gives us the upper hand. Everything went as planned and we were able to assure them safe passage through our AO.

I had planned to get on the web and chat with Roberta at 0130, right after my flight, but I didn't make it back until about 0200. I did make my way down to the internet café after flying and was still able to get online and track her down. She had been down at some friends visiting. I chatted with my mother-in-law and brother-in-law for a bit before talking to her.

8 June 2004

I went to breakfast about 0530. The chow hall has been out of butter and soda for a long time. I'm not much of a soda drinker but I didn't realize the lack of butter with my breakfast could make such a difference. Each day I walk in hoping that today will be the day it has been restocked. Something so simple can make my day better. When we're flying in support of the food resupply convoy, we joke about it being the convoy with butter and soda in it. Let's get these boys to our FOB quickly and safely! I imagine it is and it makes the mission more entertaining.

After breakfast, I went back to the hangar and went to sleep for a while before waking up and showering. After my shower, I couldn't sleep so I watched Pirates of the Caribbean to pass the time. The dry climate has caused me to have some issues with my sinuses drying out and my nose to bleed. I didn't realize that's what it was until used a flashlight to inspect it in the mirror. I felt as if I had dirt in my nose no matter how much I clean it. The inside of my nose was so dried out that the skin had split and it was cracking open. I've started using medicated lip balm on it and it helped.

I don't know if the flight schedule will be affected, but we have several aircraft with problems right now. We may have to cut the flight schedule until our maintenance group gets them repaired. I wouldn't mind, I'm ready for a break. I've been here about five months and have flown almost 200 hours, which is nearly a year and a half worth of flying back home.

Roberta is still in Missouri enjoying herself. My brother will be going to Missouri for a visit soon so it would be nice to coordinate our trips at the same time. I look forward to visiting with he and my dad. This deployment has changed me. I'll be different toward people. I'd say the biggest change will be with my kids. I don't know if I'll be able to punish them for a while, just because of the affect this trip and environment has had on me. I miss them tremendously. I know they feel the same way.

Overall, has been a great growing experience. I pray to learn all God has for me while I'm serving. We become closer to God in times of distress. We need Him most when nothing seems to be going right. I still remember having the sickest feeling ever when I heard that we were coming back into Iraq.

9 June 2004

I took a shower after my flight shift was over and then read a chapter in one of my books before racking out. I didn't wake up until about 1300 hours. Enos, a high school friend, sent me an email today. I haven't heard from him in a very long time. He is doing well. I enjoyed the memories we made together. His kids are growing up like mine. Back then, it seemed like life was never going to start for us but now it's quite literally "flying by." Many of my desires for this physical life have come true. I always prayed that our financial needs would be met and they are. My perception of enough wealth was having all our basic needs met and a little more. It's frustrating to want for basic necessities. We were so poor, we had to finance our auto repairs. I hated having to do that. My family deserves better and now they have it. Military pay won't make you rich but you are taken care of.

I've thought about finances in a different way since being in the military. I enjoy the security of not worrying about being fired because of some screwball boss. Scheduled promotions based on time and performance is the best. Overall, I'm happy where my career is.

I need to inventory my bags tonight and document it for packing and insurance purposes. I have QRF tonight so it will be a prime opportunity to knock out some busy work.

We just received word that our chow hall is out of food. We'll be eating MREs until they are resupplied. Now there's something to look forward to.

10 June 2004

I woke up to an explosion rattling the hangar. My heart was pounding but it took me only seconds to realize that it was our EOD unit blowing up weapons collected from enemy caches. The explosions shake our building so hard that the doors swing open. We are in a sheet metal building and when the blasts occur, some of the metal vibrates like a crack of lightning. It's startling for a few seconds. This is the worst part of being on night shift. The EOD teams only conduct these controlled blasts during the day which isn't too conducive to my sleep schedule. I'm flying tonight so I better get ready.

During my flight, we heard a call over the radio that the chow hall received their resupply so we are back in business! That's a motivator for everybody. We were only out of chow for a day. The flights feel longer every day. My left leg continues to bother me on the longer flights and is numbing quicker each day. We shouldn't be flying much longer. Hopefully they'll cut missions in the near future. Everyone is on edge and ready to go home. I think some people are on the verge of going nuts. It doesn't help that there are spouses cheating back home. It makes them ineffective over here. They may as well not be here. We also had a small scuffle between two pilots over something pointless. All are effects of being here too long. Some have been here nearly fifteen months now.

11 June 2004

I went to breakfast and it was fantastic. After breakfast I went for a run. That probably wasn't the best order to accomplish them so I took it slow. We'll soon have a squadron run and a spur ride before leaving so it's probably best to run now and then. We should all be getting our combat spurs at some point. The spur ride is a physically challenging event that new Cavalry Troopers participate in to earn their spurs. This is usually done back in the States and is meant to take the place of the rigors of actual combat. When you deploy to combat, you should automatically earn your spurs so I'm not sure why we're doing a spur ride in combat. It has everyone scratching their heads. Once the spur ride is complete, everyone will attend a ceremony for the donning of the spur. Silver spurs are awarded stateside and brass is awarded in combat. Our unit didn't have brass, so although in combat we will be given silver spurs.

After running, I took my shower and went to bed. I didn't wake up until 1500 hours, which is good because I've been waking up at 1300 which equated to me being up for 17-18 hours and sleeping only about 5. The routine gets exhausting day after day. It's time for some chow. Not much else happened this evening. We're now watching the Band of Brothers series to pass the time. I'd say it's one of the best series I've seen and certainly appropriate under the circumstances.

12 June 2004

I was off tonight and didn't have to fly. It was a fairly uneventful day. EOD is still detonating rounds daily and close to our FOB. The first explosion is the worst. Everyone freezes and then look at their watches. EOD blasts are scheduled at specific times so we'll know whether it's enemy or friendly fire. I played basketball with the guys this morning using a make shift goal. I'm sure I'll be sore from this.

On the war front, we're slowly pulling the forces back to see how well the Iraqis function without us. The Iraqi soldiers still have confidence issues. When our ground forces are with them they are confident and move with authority but if they are solo, many times will retreat and be defeated. When they're on their own, it's a different story. There's an Iraqi Basic Training camp on our FOB, not far from the air field. Its small scale compared to our training. They need a lesson in values. Things like loyalty and integrity don't come easy. They'll sell their friend out for five bucks.

13 June 2004

I sent a couple emails and we packed up the mil-van. It was a nice feeling to get that thing loaded. We'll still have to unload it again for the U.S. Customs inspection. This is all part of the suck of being in the military. Unload it; load it up again. Arrrgh, what a pain! It does however mean that we are going home soon so I enjoy seeing progress in that respect.

I slept in until the evening but got up in time for supper chow before departing for our mission around 2100. We completed our mission in time for midnight chow. It's usually leftovers from other meals, but it's still hot food after a long hot mission. The coffee is not bad so I drink a lot of it. I found out that the way the flight schedule was put together will cause my crew to have to fly several days in a row and the others to be off. Oh well, the missions are getting shorter and more bearable. The issue with my left leg going to sleep continues to get worse. The UH60 seat cushion has helped tremendously. The cushion delays the onset of the numbness. Anything I can find to help is welcomed at this point.

We finished Band of Brothers tonight. It's a long series but is based on a true story which makes it far more interesting. Although not the greatest, our accommodations are much better than theirs were. They watched their fellow soldiers die one by one. Wars are fought much differently now. I've not lost any close friends since getting here and hope I don't. Lord, I pray for your protection as we are flying missions. Keep us safe so we may return to our families.

14 June 2004

The air conditioning is out in our sleep room so I couldn't sleep. The electric went off and the temperature is around 110 today. At that level, the sleep room is hot in only a few minutes. I checked email this morning and received a message from Bertie. I enjoy hearing from her and always have a little something to say to her as well. I'll be home soon sweetie. I love you. Those two lines will always make it in there. I am flying again tonight. With another movie series down, I don't know what we'll watch after our missions are complete. We completed another route recon and security mission. We're checking them off one by one and counting the days until we leave.

15 June 2004

I'm flying again tonight so I tried to sleep a little longer today since the air was out most of the day yesterday. I seem to sleep longer every day. I'm now getting a solid six or seven hours of sleep daily. I normally get about five hours sleep before waking up to an EOD blast, then it's back to sleep for another hour or so. I went to chow, came back to brief and pre-flight, and then launched for our mission. The night was uneventful as we buzzed peacefully through the sky. After our return, we sat around the CP bs-ing about how nice it will be to get back home. It's finally my favorite meal of the day, breakfast.

16 June 2004

I went for a run again this morning and ran about 4.5 miles. I'm in great shape for the squadron run. We have some other troopers coming to Al-Kut for the spur ride on the same day. I didn't fall asleep until around 1030 and then woke up at the first EOD blast at 1230. I went back to sleep and didn't wake up until 1700. After chow, we flew our mission. While conducting a route recon, we saw a barrage of tracer fire pointed in our general direction. We were never able to identify the source since it only occurred once. Since we fly completely blacked out, they may have been firing in our direction hoping to hit something. We reported the incident and continued down the route. The remainder of the flight went smooth and without incident. I enjoy the adrenaline rush when something does happen. It breaks up the monotony of the daily grind.

After completing our mission, the Commander and I listened to a Ron White CD. Comic relief is a must. No one here is short on sarcasm, that's for sure. We continuously look for something to criticize and throw some unruly sarcasm at.

17 June 2004

I called home early this morning and talked with my babies. The girls are doing well, as is Roberta. She is going on a well-deserved trip. I hope she enjoys herself. I went for a run this morning after eating and that wasn't such a good plan. About 3.5 miles into it, I was getting rather sick. I slowed down to avoid puking. This running schedule just isn't working out with my daily planned activities. Maybe I shouldn't run anymore. I can easily handle this squadron run.

18 June 2004

Our mission tonight was a short and not too strenuous. I'm still racking up the flight hours. The mission went well and we made it back safely. It's difficult to get continuous rest during the day from all the EOD blasts. Again, the first one always gets my heart pounding. The inside of the building shakes wildly with each blast. It makes for a nice size dust cloud too.

19 June 2004

Everyday seems like groundhog day but none the less, they're slowly disappearing from the calendar and getting us closer to our departure. Nomad Troop will leave Babylon and re-join us today. I slept until about 1700 and when I woke up, Nomad Troop had already arrived. I went to chow with one of their pilots and caught up on events from the last few months. We're packing stuff up now and preparing to get out of here.

After several hours of packing festivities, we went to midnight chow and had the ribs. Surprisingly, they were good enough for me to go back for seconds. I'm not sure if the food is actually better or if I'm growing accustomed to dog food. I skipped breakfast this morning and went for a run. After my long slow run, I showered and went to breakfast. Surprisingly, I wasn't tired, even after staying up all night, so I read a book in the CP for a while. I finally went to sleep and didn't wake up until 1830 when my Lieutenant came in and threw a shirt on my head. I suffer because of his boredom. I sprang from my cot and out into the sweltering heat. The days continue to get hotter.

20 June 2004

We flew a mission tonight and it was uneventful. We start a new schedule tomorrow so I'll be off and have no obligations until Tuesday night. I'm counting the days down now. I can't wait to see my wife. I miss her more each day.

The heat is getting more extreme. I had some water in the freezer for two days and went to get it early this evening around 1900 and it was still water. During the night it will freeze but in the extreme heat of the day, the refrigerator can't keep up. The refrigerator warms to a cool 80F in the hottest part of the day. Sometimes I wonder if the refrigerator is working. I was able to call home early yesterday morning but the connection was so bad, it disconnected. The average high right now is 44 and low is 30. Oh, that's Celsius by the way, which equates to 111F during the day and a cool 86F at night. And it's getting hotter each day. We'll see 115-120F before we leave this sandbox.

21 June 2004

It's 0140 in the morning and I'm checking my email. I was happy to see some email from home. My little girls sent me a Father's Day email - what a treat. Naelyn has grown up a lot since I've been gone. She's reading on a much higher level. She will be testing for the gifted/talented program. Maddy seems to be having a great time at the new house and slowly making friends.

I thought I was off the schedule today but they woke me up for a mission. I had previously checked the mission schedule and wasn't on it. I love how they edit the schedule and don't put the information out, but get upset when you don't show up. Irritating! I am ready to get away from everyone for a while.

22 June 2004

I checked email early this morning again and had another email from Naelyn. She is such a cutie. Madison hasn't sent another one yet but I'll be waiting patiently. Our flight team watched a movie to kill some time, and then I went outside on the parking ramp and prayed for quite some time. It's so peaceful and cool on the airfield at night. It was refreshing.

I'm very aggravated at another pilot right now. If I show up a little early and do most of the work in preparation for the mission, he shows up even later the next day. It doesn't pay to be proactive so I'll show up at brief time like he does. He's the pilot in command and should be setting the example; let's go with his example for a while and see how things go.

23 June 2004

I spent some more time under the night sky praying for safety and that our transition home would be smooth. My foul mood slowly faded with the morning sunrise. We ate breakfast at 0500 and I dropped some laundry this morning before heading back to the hangar.

I only slept for 4.5 hours before waking up. I watched a movie with a handful of our guys, followed by some light reading, and then took a nap before dinner chow. After the mile and a half walk to the chow hall, we joined the long line to get inside. While we were standing in line for supper, suddenly.....BANG! A shot was fired only five feet from me. I flinched and did a quick 360 degree scan. A Ukrainian soldier had fired his weapon. He was clearing his weapon at the clearing barrel and failed to do one important step, TAKE OUT THE MAGAZINE! This is one way to quickly draw attention to yourself. A clearing barrel is a fifty-five gallon barrel with sand in it that allows soldiers to verify their weapons are clear or safe prior to entering the chow tent. These barrels are placed in several areas. They have them at every entrance on the FOB and virtually every building.

Our unit is conducting an awards ceremony in the hangar this evening. Rumor has it that we are cut off on awards because too many were given out. Something doesn't sound right about that but its reality. Tomorrow holds more festivities - the spur ride. I don't fully understand the purpose of a spur ride in combat but like everything else in the military, I'll do as I'm told.

We just finished a troop formation where the SCO and SGM presented everyone with squadron coins for Operation Iraqi Freedom. I'm proud to be part of this unit. It's only a coin but it's also a constant reminder of our mission and what we've accomplished. I guess I should study the Regimental song tonight for the spur ride tomorrow as I'm sure I'll have the displeasure of singing to someone before the sun sets again.

24 June 2004

I was on the QRF mission today so didn't have to make the squadron run. I did however, participate in the spur ride and dinner. The spur ride is an event where military rank is erased and the only authority lies with the wearer of the spurs. Without spurs, you have no rank. I made it through but was very sore. I've been up for the better part of 30 hours. We had a ceremony for the presentation and donning of the combat spurs on each new spur holder, followed by a dinner of celebration. After the spur dinner, I didn't want to go to bed. It would have thrown off my sleep schedule, so I stayed up most of the night with only a short nap around 0100 hours.

I'm going to the shooting range tomorrow to shoot some small arms weapons with a few pilots. We have some ammo to expend so we're going to have a little fun. We'll grab breakfast at 0500 before heading over there. With us so close to leaving Iraq, I've been in a great mood.

25 June 2004

We are preparing equipment for departure. My sleep schedule is really screwy right now. I'm only able to sleep four hours at a time. Today, the temperature reached 109F and was a little more humid than normal. The moon is slowly showing itself at night again. With no moon, Iraq is a very dark place at night. I enjoy praying at night on the flight line. It's very peaceful most of the time. All of our crews have been working on cleaning aircraft for the trip home.

We made the trip to the small arms range this morning and had the pleasure of shooting with the Ukrainians. They brought their BRDM patrol vehicle and fired the gun on it. That was a treat. We expended several thousand rounds of ammo from many types of weapons. Shooting an AK-47 on full auto was a new experience. It only takes a few seconds to empty a thirty round mag in full auto mode. What a rush! We also fired a full auto, mafia style, grease gun. It could empty a clip in about 1.5 seconds.

26 June 2004

Today was another groundhog day, more of the same routine. I did get a chance to email home and also did some web research on financial investing. I'm really bored so I'm desperately trying to keep my mind active. I'll do some reading and more packing today. I have some worn out clothes that need to be tossed. In the past, some of our old uniforms have fallen into the wrong hands, which proved to be deadly for coalition forces, so now we are burning all our worn out uniforms. Packing won't take long because I'm only packing the things I won't use between now and when we pull stakes to leave.

27 June 2004

I marked another day off of the calendar. It feels nice to put a line through another day. As I went out to the port-a-john in the middle of the night, I realized I've memorized getting to and from it without the aid of even a flashlight. I had a conversation with a fellow soldier today about how tired I was of walking or riding in the back of a Humvee a mile and a half just to eat each meal. I'm ready to go home. I can't wait to see my girls.

28 June 2004

I was able to sleep last night so I should be able to reverse my sleep schedule relatively easily over the next couple days. The time has come to turn this area over to the Iraqi people so the Iraqi leadership conducted their transfer of authority (TOA) today. Originally announced to be conducted on 1 July, it was completed today to minimize enemy activity during the transition. Tactically this is a great plan since it's not expected by the enemy insurgency. The goal of the Sadr militia is to upset the Iraqi people during the transition in an attempt to inject instability into the communities and create fear, doubt, and intimidation.

I watched movies until about 1615 hours, when they came and took our TV away to pack it for the trip home. All this packing has given me a short timer's attitude. We leave in two days and I'm ready to go. I'll be happy to cross that berm between Iraq and Kuwait again. I pray nothing stops us this time; I'm ready to see my family again.

29 – 30 June 2004

We spent the last two days in June packing our goods for the short flight south into Kuwait. Spirits are high once again but there's still an underlying nervousness in the squadron - a fear that we could be extended again. I remain positive and focus on today. With everything packed but a small "go bag," I bedded down for what I hope is my last night in Al-Kut.

1 -13 July 2004

Our last night in Al-Kut was a quiet one. I slept well and was ready to fly this morning. We readied the aircraft and completed the mission briefing last night, so after breakfast all we had to do was crank 'em up and fly out. Like a flock of geese flying south for winter, we wasted no time getting airborne. Our next stop was Kuwait.

After crossing the border, we landed in Udahri and de-armed the aircraft weapon systems before parking them for the night. We began the redeployment routine that was very familiar, having done it only three months ago.

14 -15 July 2004

Feeling very anxious about the return trip home, it had finally arrived. I felt like an eight year old waiting for Christmas morning. Weapons in hand, over 300 of us boarded an Omni Airlines jet headed home. When we lifted off the runway, joyous clapping broke out throughout the airplane. A tension lifted and we were finally homebound.

After six hours in the air, our first stop was in Germany. We were allowed to exit the airplane while it was serviced. The terminal had several shops with neat trinkets, candy, and souvenirs. All of it was expensive but I splurged and indulged myself with some German chocolate. We boarded the plane once again for an eight hour leg to Maine. Once again, we clapped and yelled in joy when the tires made that glorious bark from contacting U.S. soil. When we deplaned, we were met by many local supporters that lined up along the skywalk. As we entered the terminal, they began to clap and thank us for our service. Their gratitude sparked an unbelievably proud feeling inside me. I was ten feet tall and bullet proof. That's a moment I'll never forget. They handed out several cell phones to allow us to call our families and let them know we had safely arrived back in the U.S.

After a short break, we re-boarded the plane for the last time. Our next stop was Alexandria, LA, just east of Ft. Polk, where we were based. After landing in Alexandria, we loaded up in several tour buses to make the one and a half hour drive to the post. Our families had gathered at a gymnasium in Fort Polk, where they waited for us over two hours. The closer we got to home, the more anxious I was. I would finally be reunited with my wife and daughters. After arrival, we filed off the buses one by one and into the gym where we fell into formation for one last time before being dismissed to our families. The Regimental Commander spoke a few words then dismissed us.

The bleachers were full of family and friends and the scene was chaotic as we scanned the crowd for our loved ones. People were screaming and tears of joy rolled down many faces. I spotted my girls waving in the stands. It had been so long since I'd seen them that it was strange in a way. The girls seemed so small. Their faces were red from the heat in the gym. I hugged and kissed my wife, then scooped up my little girls and squeezed them tight. I looked at them all and said "I missed you girls so much. Let's go, I'm ready for a cheeseburger."

Chapter 10

Reflagging 4/2 ACR to 4-6 ACS

16 July 2004 – 6 June 2007

After returning from Iraq our unit was more than ready for block leave, or vacation as it's known in the civilian world. We spent about 7 days completing administrative classes and paperwork before being released. The squadron also held an awards ceremony for several of us and I was presented with my first Air Medal. I had done nothing spectacular in my opinion but it was nice to be recognized for doing my job well in a combat environment.

We went home to Missouri to visit family and friends. I was more thankful for my life than ever before. I could finally spend as much time as I wanted with my girls. I enjoyed a fat and juicy steak along with a tall draft beer and life was good.

A few weeks later, the fun was over and I returned to my unit in Fort Polk to assume the duties as the Troop Supply Officer. We immediately began sending our aircraft to reset. The birds were all mechanically tired, worn out, and badly in need of refurbishing. The next big project was to lay out our unit equipment and determine what was good and what was in need of repair. We had many pieces of old and broken equipment. Some had been lost or trashed while other items, like tents, were torn or missing pieces. As the Supply Officer, I was responsible for turning in old equipment, ordering missing parts or pieces, and receiving new equipment. I had my hands full to say the least. My days would be long until things were back in shape.

It wasn't long after our return that all the Commanders were swapped out for a whole new group. They had done their time and it was someone else's turn to stand at the helm. We spent the next year refitting the unit with equipment and aircraft. Early in 2005, it was announced that the Armored Cavalry Regiment would be broken apart. We would no longer be 4th Squadron in the 2nd Armored Cavalry Regiment, but we would re-flag and become 4th Squadron 6th United States Cavalry, or 4-6 ACS. Part of this re-flagging would include moving the entire Squadron to Fort Lewis, Washington.

Spring came and went and our unit began moving in early summer of 2005. Fort Lewis wasn't prepared to receive our 600 plus soldiers at the time so the housing availability was slim. Roberta and I had purchased a house in Louisiana because of no availability and here we sat less than a year later with the same dilemma. We weren't in a good position to sell and the housing prices are booming so we decided to hang on to it and rent it out. Housing in Washington was expensive compared to Louisiana so we were conservative and bought a new build outside of town, hoping not to overextend ourselves.

We pulled stakes in Louisiana two months prior to hurricane Rita hitting the gulf coast. We timed our move to be just after the contractors promised to have our house completed, but the builders were far behind schedule. We lived in a hotel for an additional four weeks before it was completed. We were at each other's throats by this time and ready to get into our new house. It was a relief to finally move in. The house was a modest 1500 square foot two story with a garage. It sits in a community at the outer foothills of Mount Rainier. Situated in a nice gated community, it wasn't uncommon to see deer in the front yard or a herd of elk on the thirty minute drive to the base. We had moved to a serene and peaceful area. I felt as if it separated work from home, something we never had while living on the military base.

By now, we have received much of our new equipment and most of the helicopters have returned from reset. It was time to test it all out. Over the past year, we had several pilots transfer to other units, retire, or exit the military so we had many new faces as their replacements flooded the squadron. This also meant many folks with little to no experience. Although not a senior aviator, I had more experience than half our troop. I spent nearly six months flying in Iraq and learned much that could be built upon and shared with the junior pilots.

In the spring of 2006, we deployed the unit to the National Training Center in the Mojave Desert of California. The deployment was only four to five weeks but in that time, we recreated the combat environment in Iraq and prepared for our next deployment that was just around the corner. I was paired with CW2 Hart who was as hungry for flight time as I was. We flew as a team from the time we left Fort Lewis until we returned. Our backgrounds were similar as was our thought processes. We worked well together and both were very effective in the aircraft.

After a week of administrative briefings, local area flights, and dust landing training we started our seven day rotation of mock combat. Matt and I reversed our sleep schedules to accommodate flying the night shift and opened the rotation by flying the first troop mission that went through most of the night. The OpFor, or tactical enemy, put a high value on bringing down helicopters, as would our insurgent enemies in Iraq. Fortunately, we avoided being shot down during the training exercise. Several aircraft were downed that week but ours wasn't one of 'em. Our confidence was high, but we remained humble knowing it could all change in an instant. We flew about thirty five hours that week and nearly fifty hours during the month. Being away from home was difficult and reminded me what being deployed felt like. I dread leaving my family behind.

Several months passed and our Squadron Command changed again. LTC Jamison would be the commander to lead us to combat in the next Iraq rotation. CW4 Morris had been appointed as the Squadron Standards Officer for the Kiowa pilots. Both brought much experience and leadership to the squadron. By the time we were leaving in June of 2007, over a year would pass since our NTC rotation so it was decided that we needed another training exercise before deploying to Iraq. Our unit had grown to around 850 soldiers and would be a logistical nightmare to move this soon before deployment so the Army decided to bring the exercise to us. Elements from the Joint Readiness Training Center (JRTC) in Fort Polk, Louisiana, were deployed to Fort Lewis to accomplish the mission.

Now a more senior aviator, I was being paired with low time pilots. I was appointed to a Unit Trainer position and tasked with conducting mission training with junior pilots. After completing a seven day exercise, we flew our aircraft to the port at Grays Harbor. The birds were loaded on a ship and would soon be under-way to Kuwait.

There were so many war protestors in Washington that we had to change the date of the aircraft movement to avoid mob crowds of protestors at the port. Upon arrival at the port, I saw two layers of security between the entrance and the landing area. Changing the date from the previously published date was a success. There were only a handful people standing outside the fence in protest. Protesting wasn't uncommon at the Fort Lewis gates. The north I-5 entrance was known for strong protesting while the south I-5 entrance was crowded with supporters. I had mixed feelings about this but after a few months it felt normal and I thought nothing of it.

At this point, we already knew that we were being deployed for fifteen months. I'd rather know up front that we'll be gone that long. We were also being told that we were most likely going to the northern part of Iraq, which had been a very hot zone in the past twelve months. This area had claimed many lives and we would soon be in the mix. With the helos tucked into bed on the ship and equipment packed, we went on block leave to see our families and friends one last time before leaving the States.

Chapter 11

Another Dreaded Goodbye

7 June 2007

The whole family was up early this morning in preparation for our deployment to Iraq. This most certainly proved to be one of the toughest days of the deployment. Saying goodbye is never easy, mostly when there's a great deal of uncertainty. We met in our hangar several hours before the buses were scheduled to depart. Having flown the aircraft to port weeks earlier, the empty hangar echoed every sound. Weapons and night vision goggles littered the floor and they slowly disappeared as we filed through each station and secured our assigned equipment. Our family members tagged along and tried to hide the emotion as if it were just another day of deploying to a field exercise. I didn't want them to worry so I played it off to my daughters as if it was no big deal and said "I'll be home before you know it." We sat quietly for a few minutes, then heard the roar of the buses lining up outside. I had a sinking feeling in my stomach. It was the anxiety of voluntarily walking away from my girls and thinking this could be the last time I see them. I had a few tears from the emotion and said a short prayer asking God to protect us all before saying a final goodbye and boarding the bus.

Personal gear ready for loading

Family shot

Last goodbyes

Buses used to transport us from Fort Lewis to our plane
at McChord Air Force Base

As we rode away on the buses, I waved to my girls and did my best not to dwell on the worst outcome. As always, we shifted our thoughts to sarcasm and jokes to cope with the family separation anxiety. After about an hour wait, we departed McChord Air Force Base mid-day and flew to the east coast for a fuel stop before crossing the Atlantic. The plane was crowded and uncomfortable which made the eight hour flight to Germany a very long ride. I spent much of my time standing in the back of the plane drinking coffee and visiting with the flight attendants and other soldiers. After landing in Germany, we deplaned and spent just over an hour refuelling before boarding and departing once again.

8 June 2007

We flew six more hours in the sardine can before landing in Kuwait around 1900 local time. Dang, it was hot! It's a culture shock leaving Washington State where the summer highs have been 75° and stepping off the plane into a balmy 112° in the shade. I have forgotten how bad the Kuwait heat is. I felt like I stepped into an oven. The environment is harsh but my determination is greater and I'm overcoming it mentally. The key to overcoming it is to see yourself as a machine, as if nothing can bring you down. If you overcome it in your mind, you can overcome in reality.

After all the briefings and baggage claim, I dropped my gear in my sleep tent mid-morning on the 9th. I managed to call home and let the girls know I arrived safely and wished Roberta a Happy Birthday. We've been up for over 48 hours and we're all exhausted and ready for some sleep.

9 June 2007

I felt rotten most of the day and attempted to sleep through the heat. Unfortunately, I haven't been able to. It's unbearable. The air conditioners in the tent are screaming, but still aren't keeping up and it's too hot to sleep. The high today was 122°. We had two large air units for our tent and the air inside is still 98°. After realizing that I wouldn't be able to sleep, I wondered around the camp to see if I could find a cooler spot and familiarize myself with some of the amenities. There were several nice little shops and fast food options. The problem with most is that they're in portable trailers and don't have inside seating. There is an ice cream and coffee shop which seems to be the best place to kill time because of their air conditioning.

10 June 2007

Temps are cooler today with highs hitting only 107°. I heard there's a shamal coming within the next day or so. I equate being in a shamal to standing in a sandblaster. The wind is already picking up and visibility is dropping. The port operations and maintenance teams are unpacking the helos, so we should be flying soon. The sooner the better because flying makes the time go by much quicker. I attempted to call home but the call timed out as it rang into an empty house.

11 June 2007

Our Troop Commander asked me to prepare a finance briefing to discuss financial responsibility and investment options with our troop. During deployment, we're paid a few more dollars than usual. On average, I'd guess it's about $500 per month more. Many families will develop financial problems while we're deployed. Some spouses will unintentionally use spending as a coping mechanism while their soldier is away. Separation anxiety is a mental stress that affects everyone differently and if not prepared for, can completely tear a family apart. I know of several families that it's affected. Some survived while others didn't. These are the hardships of a deployment. Roberta and I discussed this and setup automatic allotments to keep the extra money out of sight and mind. The more issues you can keep at bay, the better. Keeping your personal affairs in order at home directly relates to effectiveness in combat.

I called home and heard that my oldest daughter Naelyn was awarded the Academic Achievement Award from her school. I'm extremely proud of our girls and hated that I missed the end of year ceremony.

12 June 2007

I stayed up late last night in an attempt to sleep longer this morning. That was a fail. I woke up at 0600. We had a few mandatory classes today dealing with things like Rules of Engagement, better known as ROE. I managed to take a three hour nap and felt much better after waking up. I have several more hours of free time, which isn't good because I think back to the day we left and the look on my girls' faces. We've only been gone a few days and I already think about going home. I hope the girls don't dwell on it and just let the time pass. I should take my own advice.

I usually spend a little time each day in prayer to reset my mind. I believe God has me here for a purpose so I gladly accept the challenge and will drive on. I enjoy my military job thoroughly. Although dangerous, this will be a life-changing experience in the end.

The best way to cope with the hard times is to extract the good things and stay focused on what motivates me. This is something I'm good at. I need to hit the sack early tonight since we're getting up at 0300 to pick up the helos from the port tomorrow. I'm slowly acclimating to the environment and new time zone.

13 June 2007

We've been here four days doing nearly nothing and it's time to make the donuts. We rolled out of our racks at 0300 and piled into a bus, bound for the port. I'm not sure how far away the port is, but I do remember them briefing that we weren't allowed to drive through Kuwait and we would have to make the long trek around the city. Each bus had a "point man" that would be assigned three rounds for his M-9. Yes you read that correctly, three rounds. I'm not sure why only three rounds, but we always joked that it was so you could shoot twice and if you missed, you would have one left over for yourself.

We weaved through multiple security checkpoints leaving the camp and entering the port area. It was five and a half hours before we ever saw our birds. The departure process was slow and the sun was getting hotter. It was 112° in the shade and 130° in the sun. The scorching asphalt made the air dance in the distance. I drank several bottles of water while we waited under a screen shade. The wind began to pick up and only a few teams departed before a wind warning was issued and we were restricted from flying. The remaining teams, including mine, would remain overnight and try it again tomorrow.

Several of us piled into a small white mini-van and headed to Camp Arifjan in Kuwait. Although only fifteen miles from the port, it took us an hour and a half to get to our billeting area. I felt sick as soon as we left the port. We were packed into the van shoulder to shoulder and I was getting heat exhausted. The van would overheat if you ran the air conditioner in stopped traffic so the driver had it shut off the majority of the time. We were stuck in a long line of cars that were waiting to pass the gauntlet of the security check point to enter the camp. I had a splitting headache and felt nauseated. The heat in the van was getting worse and I was heat exhausted. I finally said "I have to get out of this van." We were about 100 yards from the checkpoint so a few of the guys got out with me and we walked to the check point to ask for water. They gave us several bottles of water. I immediately loosened my clothes, unlaced my boots, and removed my shirt. They grabbed several bottles of water and poured them over my head as I drank more water. It helped quite a bit. We continued this process until our van was searched and cleared through the check point.

Most of the guys were happy to stay at Camp Arifjan because it has a swimming pool. I was happy to see a bed in an air conditioned room. We toted our bags up to our bunk bay and I immediately took my boots off and went to bed. The rest of the crew bolted to the PX to buy swim trunks and hit the pool. As bad as I wanted to get into the refreshing water, I couldn't, I felt like death. A cool bed was my oasis and I was counting sheep before they left the room.

After a three hour nap, I woke up feeling refreshed and headed to chow. It was very apparent that Camp Arifjan was an Air Force run facility. There is some serious money being spent here. The chow hall served steak and lobster for dinner. Wow! What a treat that was. I feasted like a king then high-tailed it over to the PX to pick-up some basic overnight items for the unplanned stay. I made it back to the room around 2000, showered, and hit the sack. I slept like a baby on one of the worst mattresses I've ever laid on.

14 June 2007

We woke up to our alarms at 0300 again. This early start should get us out of port before the high winds develop from the mid-day heat. I felt like a new man this morning. The water, cool air, and extra sleep recharged my batteries. We spent a few minutes taking another look at the birds before cranking up and pulling pitch. As we lifted from the port, we had a good view of Kuwait. It's a dusty and dirty place. Beyond the city, there is absolutely nothing but dirt. Why would anyone want to live here? It's so desolate. We flew over camel herders along our route back to the airfield. Strangely enough, we also saw a camel track equivalent to our horse racing tracks. Camels are not graceful animals and it's comical to imagine them running a race. I'm sure the camel jockeys would disagree.

We arrived at the airfield just in time to miss breakfast chow, so I went to the Subway trailer and bought a sandwich. We completed the main task of the day and it was only 1000 hours. I picked up my laundry from the cleaners and signed up for a ten minute time slot for the internet over at the USO tent. The tent was full, but I had nothing better to do and they have air conditioning. We're scheduled to shoot our small arms at the range tomorrow. I'm ready to get this party started.

15 June 2007

Our troop had an early slot for the range so we got up at 0400. The range ops were run by an ex-delta force guy and reminded me of the range that Lieutenant Rosnick set up back home. I shot well and validated my M-9 was on target and ready for a challenge.

After returning, I went to lunch with a few of our pilots and then took my computer to the USO in hopes of seeing the girls over the web cam. I called Roberta and woke her and she in turn woke the girls. It was nice seeing their faces. We spent two hours chatting online and it boosted my spirit. I've only been gone eight days and it seems like forever.

The generator that powers the air conditioners for our sleep tent went out and it only took a few minutes for it to heat up to triple digits inside. We're living the tent life right now. The cots are only separated by a few feet so it's very cramped living quarters. It's much worse when the air is out. It got very smelly really fast in the tent. A maintenance crew was called and it was repaired in short order. After the repair it still took a long time to get the temps back down to a bearable level, in the 90's.

16 June 2007

We were supposed to fly out to the test fire range today and test the aircraft weapons systems but the winds were once again too high. With nothing else on my schedule, I called home and talked to the girls for a few minutes. I attempted to contact my parents but was unable to reach them. I'm not sure what opportunities I'll have to talk to everyone back home once we're in Iraq so I call every chance I get. I'm going to bed early tonight so I'll be well rested.

17 June 2007

Today is Father's Day. I got up at 0600 and went to breakfast before flying. We checked the aircraft weapons and mission equipment in preparation for crossing into Iraq. I enjoyed a game of chess with another pilot midday. I ran into a pilot from my flight school class while at the chow hall. We shared experiences from our past deployments and enjoyed the time catching up on the past few years.

Andy and I took our computers to the USO and I made a web call to Roberta. She was there waiting with her morning coffee and a smile. The USO randomly handed out calling cards so I used mine to call my dad and wish him a happy Father's Day. I returned to the sleep tent late in the day, showered, and hit the sack.

18 June 2007

I had to attend a class on how to setup the network hardware for a supply system we would be using while in Iraq. I am the Supply Officer for our troop and take care of ordering new equipment, replacing broken equipment, and maintaining accountability for everything on the books.

Our air conditioner in the tent is working better but it's still hot. I spend most of my free time at the USO tent. The tight living quarters and hot tents have everyone very irritable. We're all ready to move north into Iraq and get into our mission. So far it feels like another exercise. I'm sure it will be very real when we get into Iraq and the bullets start flying.

We flew our night environmental flights tonight. This is a flight that we complete with an instructor pilot to train and practice landing in the worst environmental conditions in the area, at night while using night vision goggles (NVGs). The NVGs use ambient light, such as moon light, to function but we had no moon illumination so therefore little effectiveness compared to a moonlit night. We flew into the desert well away from civilization and practiced landing in the thick powdered dust on the desert floor. This landing is one of the most challenging landings I've experienced. As you approach the ground, the powder engulfs the aircraft and can cause you to lose your bearing. The doors are removed on the Kiowa to allow us more visibility from the cockpit and so we can easily use other weapons such as our M-4 or to throw grenades. On the down side, it also allows dirt to be blown inside the aircraft during this type of landing. Each landing can fill your nose, mouth, and eyes with dirt so it has to be completed using good crew coordination and communication between the pilot and the co-pilot. Needless to say, I needed a shower after the flight.

19 June 2007

I attended the second half of the computer class to learn to setup and operate the supply computer network infrastructure. We finished around 1600 so it burned most of the day away. At lunch, I had a conversation with a fellow pilot in my troop about the possibility of losing one of our guys in combat and how it may affect us or the troop. Our troop is very tightly knit. I hate to think it but it could happen. We are about to enter one of the two hottest operational areas for enemy activity in Iraq and this may be something we have to face. Losing a soldier is the elephant in the room, a topic that no one wants to talk about and we pray it never happens.

I often pray for the safety of not only myself but also our troop and squadron. I recite Psalms 23:4. "Yea though I walk through the valley of death, I will fear no evil, for you are with me." I asked all my family and friends back home to remember our unit in their daily prayers as well.

20 June 2007

I was tasked as part of a team to conduct an aerial range sweep this morning. A range sweep is completed to verify there are no people or animals on the firing range before we go "hot" and begin test firing. While flying to the range, we saw a herd of approximately 100 camels with two shepherds. I snapped a few photos to send home. The kids will think that's pretty neat.

Although still early, it was already very hot. I'm not sure what the temperature was but I was soaked in sweat. We flew back to the airfield, landed, and shut the birds down. I had the rest of the day to kill time so I worked on building a troop web page so we could share photos with our families back home.

21 June 2007

I flew with another team today. We'll continue rotating our birds through the range until all the aircraft weapon systems have been fully verified. Today was another scorching day. It was between 110°-115°. It feels even hotter with the flight suit and body armor on.

While out verifying the weapons, we had a problem with the aircraft so we flew it back to the refueling point and shut it down. The engine oil temperature was too high. This turned out to be a long and hot day for me. We called for the maintenance team to come look at it. It took them a couple hours to get out there and they brought one of our maintenance pilots with them. I flew along with the maintenance pilot and we babied it back to the airfield so they could work on it.

After gathering all my gear, I went over to the USO tent to cool down and try to get some internet time with my girls back home. I was able to chat with them for a while and it was the highlight of my day. My batteries are charged.

22 June 2007

We're slowly getting all the mandatory items completed so we can move into Iraq. With so many moving parts to our large squadron, it takes time to make sure each FOB is also ready for us in Iraq. Our squadron is splitting up and will eventually settle at four different FOBs in the northern part of Iraq.

Today began a series of squadron level briefings on the threat environment, our mission, and our individual areas of operation. Our squadron is comprised of three troops of Kiowa scout helicopters (Ace, Blackdeath, and Carnage) and one troop of Blackhawk helicopters (Darkhorse). Ace Troop has been assigned to the city of Mosul and would be co-located with the squadron Tactical Operations Center also known as the TOC. Our troop, Blackdeath, is going to FOB Sykes, near Tal-Afar, located west of Mosul and sits just south of the Syrian border. Carnage Troop will be on FOB Speicher, near Tikrit, roughly 120 miles south of Mosul. Darkhorse Troop will be divided among the FOBs and fly mainly logistical support missions throughout the AO. A group of our Blackhawks would eventually be assigned to Kirkuk, well east of Mosul. Our squadron will be the only aviation presence in the northern part of the country for the better part of a year.

We are replacing 1-17 Cavalry so they sent one of their Tactical Operations officers to Kuwait to give us a briefing on the threat in each of our individual operations areas. Mosul undoubtedly had the greatest threat to aircraft. Aircraft are being shot at daily and the FOB is attacked often. Mosul is a big city and an easy place to shoot then meld into the population. After losing one of their own pilots in Mosul, 1-17 established "no fly" zones within the city because of a heavy enemy presence. As scary as these areas are, we can't have no fly zones. We must break the enemy. A no fly zone says, "you win." I expect our Squadron Commander will eliminate those and immediately let them know the rules have changed. They need to know there's a new sheriff in town. Ace Troop's mission was to provide aerial security for the city, FOB, and our ground forces and to find and eliminate enemy forces within the city. I believe the population of Mosul is approximately 1.7 million.

Tal-Afar is a small town about 45 miles west of Mosul. The town had been through a very intense battle the year prior to 1-17 Cav arriving and had some very hostile people in it at one time. 4-3 Cav lost a pilot over Tal-Afar in Aug of 2005. Story was, he was engaged by small arms fire and he took a round to the head. I believe his co-pilot was also shot in the leg but managed to land the aircraft. His information suggested that Tal-Afar has been peaceful in the recent past but complacency kills so we'll watch our backs. The entire north-western area is being used to smuggle weapons and Al-Qaeda recruits into Mosul. Our mission is to provide security, recon, and air support to ground forces operating in our vast north-western territory. Our AO was several thousand square miles. With an extremely large area to cover, this will be no easy task.

Carnage Troop has a mission very similar to ours, only further south. I wasn't in the briefing for their AO so I have very few details on it. I know that they have several pockets of heavy enemy activity. I suspect there is much weapons trafficking between Baghdad and Kirkuk. It's certainly not a friendly area south of Kirkuk.

The briefing was sobering and reminded us all that this isn't a game, exercise, or another gunnery range shoot. This is real and people will be trying be trying to kill us daily. This mission is nothing like my first deployment. During my first deployment, our enemies were only beginning to organize. Toward the end of our tour, they had become somewhat organized in areas but nothing like we are facing now. I admit, I'm nervous about what we're getting into. Many of our pilots are new. This will be their first combat deployment. As a mentor to some, I have to keep my head on straight, make smart decisions, and continue to train junior pilots to employ the aircraft effectively against our enemy.

23 June 2007

We rolled out of our racks early today and headed to breakfast chow. Shortly after, we all met for another briefing. Yesterday's briefing by the out-going unit was general information and had my senses at their peak. Today our Squadron Commander met with each troop to brief our mission in greater detail and give us an idea on how each FOB is equipped.

From his briefing, I'll be living fat compared to my last tour in 2004. I hear the chow is great and it's a comfortable place as FOBs go. There's a building called the multi-use facility or "The MUF" as it's been ironically labeled. The MUF has all types of games such as pool, ping pong, card tables, a movie room, and even a table top shuffle board game. Another section of the building has a basketball and volleyball court. They have another room off of the game room that has several PCs with internet access. This place sounds better than back home. There's a full Post Exchange or PX on the FOB. We'll be living in Container Housing Units or better known as CHUs. I was relieved to hear that the pilots will be assigned individual units after 1-17 Cav is gone.

After hearing the Colonel's speech, I was motivated and ready to get up there and get to work. We're scheduled to leave in three days. Everyone is ready to get out of Kuwait. It's hot, crowded, and miserable.

24 June 2007

We had more meetings and briefings today to discuss the movement into Iraq. After the meeting, we went to the flight line to pre-flight the birds. Tomorrow we'll crank all the ships and complete systems checks for the flight out on the 26th.

After a hot day on the flight line, I spent some time at the USO tent talking to the girls back home. They're doing well.

25 June 2007

This morning's run-up drill could have been a bit smoother. There were a few hiccups but nothing major. I hope it's better tomorrow when we cross the border. Anxiety is high and tomorrow everything becomes real. The war games we played at NTC and Fort Lewis are behind us. Now the stakes will be our lives. One mistake and you could pay the ultimate price. I hope my brain was spongy enough to soak up all that training knowledge.

Our ADVON has been in Iraq for several days now. The ADVON is the advance party that moves ahead of the main body of the unit to assist in the reception of the main group of soldiers that arrive shortly behind them. Our S-3 Operations Officer is part of the ADVON team. We got word today that he was in his first engagement and shot several rockets and his M-4. There's been no word on the outcome of that engagement. The news has my blood pumping. He's only been there a few days.

I talked to the girls today and it's their last day of school. They're excited because the community pool is open every day after school is out. The daily temps back home have been around 60°, much different than our 114° in the shade today. As I lay down tonight I pray. *"Lord, protect us tomorrow as we cross into hostile territory and let us be invisible to our enemies."*

26 - 27 June 2007

We were early to rise this morning and had a good breakfast before heading to the airfield. Our troop was scheduled to depart after Ace Troop. The winds have been picking up all morning and the visibility deteriorated to a point that we were unable to launch. The weather delayed us for two days. The days are hot and the sand is blowing strong. I can't walk out of the tent without getting dust in my ears, eyes, nose, and mouth. I'm so ready to leave this dust bowl.

My back has been hurting from sleeping on the cot for so long so I went to the flight doc. He's also a D.O. so he made an adjustment on my back and prescribed me some meds for the pain. I capitalized on the down time while waiting for the weather to clear and spent a few minutes each day chatting with the girls back home. The weather forecast for tomorrow is clear so we should make it out of here.

Dressed for the dust storm

28 June 2007

0300 hours came pretty early but we were up and ready to roll. The weather looks great. Today will be the day we fly into Iraq. Ace Troop departed with the Squadron Commander leading the pack. All the ships were heavy with fuel, ammo, and personal baggage. We needed max fuel to make it to our first stop.

Ace Troop departed in teams of two. We watched as the birds lifted and turned northbound. One of their birds picked up and twisted around abnormally. We knew something wasn't right. As I watched, I was thinking, "What's going on?" The tail twisted and turned as the skids bounced off the ground. The aircraft quickly disappeared into the cloud of dust. My heart sank; this usually ends with the aircraft on its side. As the dust cleared, I could see that the aircraft had settled level and upright. They had exceeded the maximum torque. It was hot and we were heavy, a combination that helicopters don't like. Fortunately, the over-torque was minor and they were able to depart a short time later. Thankfully, we didn't have a major incident. What a way to start the trip.

Our troop cranked and departed a short time after Ace and without issue. We are in for a long hot day. Our planned flight time today is six hours. That gets us to FOB Speicher, where Carnage Troop would remain. We made our first stop without issue. Our second stop was in Al-Kut, my old stomping grounds. It was neat seeing the airfield again. The Polish are still running the airfield and greeted us in their standard uniform, shorts and t-shirts.

The winds were picking up and a dust storm had blown in just north of our current position. The visibility was still legal so we pressed on. We were only twenty minutes north of Al-Kut when the visibility decreased significantly. We had less than a quarter mile visibility. Although not legal, this was combat and we weren't landing in unknown territory. Using Blue Force Tracker or BFT, we sent a message to Ace Troop to get the weather conditions at their location north of us. They advised that the weather had cleared so we continued. Flying only by instruments, the poor visibility continued for nearly an hour before we broke out of it. I said a prayer asking God to bring us through it safely. Ironically, I was paired with an atheist for this trip. Did I mention that my number in this troop is 13 and I have a black cat? I'm not superstitious.

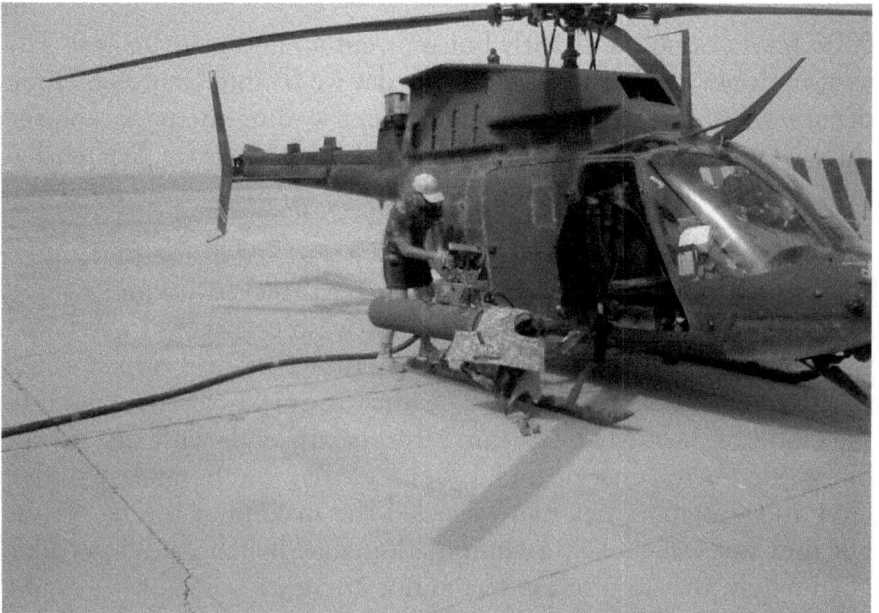

Polish soldier fueling the helo

We landed at FOB Speicher around 1330 and were more than ready to get out of the cockpit. The trip to the doc helped. My back felt good after the long flight. We were given a proper welcome by the out-going unit. They met us with cold drinks and a vehicle to transport us and our gear to our tents. They have the best chow hall I've ever seen. It's the size of a small Wal-Mart! They have a variety of selections including a main line, short order line, made to order grill, full salad bar, pizza buffet, desert bar, and any drink you could think of, including beer. It was non-alcoholic beer, but beer none the less. Things have certainly changed since 2004. Tomorrow we'll complete the last leg of our flight to our respective FOBs. The flight to Tal-Afar should take about 90 minutes. We're all ready to get settled and start our mission.

Chapter 12

Suicide Bombers

29 June 2007

We left FOB Speicher this morning and flew to FOB Sykes in Tal-Afar. It was 90 minutes of open desert and shear boredom. Again, I wonder how anything survives out here. As we approached the airfield, I did a quick visual survey of the FOB layout as we called the control tower, "Tal-Afar tower, Blackdeath 23, short final." The controller replied, "Blackdeath 23 you're cleared to land Alpha taxiway." The wide-spread airfield was lined with Kiowa Warriors. After shut down, we unloaded our personal gear, secured our weapons and NVGs, and in the great spirit of the Cavalry, mounted our Stetsons before heading to the Command Center.

A short time later, we met our Commander at the troop Command Post, better known as the CP, and received our CHU assignments. In short order, Mike and I found our CHU and dropped our gear. He and I will share a CHU for the first two weeks until the RIP / TOA is complete. RIP / TOA are acronyms for Relief in Place and Transfer of Authority. This is the period of time when we will conduct missions jointly with the outgoing unit. This process assures the best continuity for operations in the region.

Our CHU has two single beds and two small shelves. With several departing soldiers, there were many comfort items such as desks, shelves, and benches to be had. Most were for sale but if you were lucky, you could find someone giving stuff away. More bartering will take place in the next two weeks than Wall Street will see in a month. The CP had a corkboard designated specifically to sell items. I weaved through the CHUs looking for bargains and stumbled across the perfect deal. This soldier had a plastic four drawer clothes dresser, microwave, and some power cords for $10. SOLD! I gave him a twenty and told him to keep the change; it was well worth it.

After hauling my newly purchased comfort items back to my CHU, I made a trip to the PX and acquired a coffee pot. Nothing starts the morning better than a fresh cup of joe. Since the a/c power is 220v here, I waited until I was here to buy a coffee pot. Some people learned the hard way that plugging 120v electronics into a 220v outlet will make a little smoke and a lot of anger. I saw more than a few things get fried during my last tour.

During the planning phase of this trip, we heard that the troop we are replacing has their own internet system and are interested in selling it to us. Having internet directly to our individual CHUs is a huge morale booster. The system was expensive and required pre-payment from all those interested in being connected to it. Today I linked up with the pilot that is managing the equipment so I could give it a once over and see what it was capable of. The network cabling is a mess and needs replaced, but the satellite and modem are both in great condition.

We agreed to purchase the internet system and take over the monthly contract. The service is provided by a foreign company that I only have an email contact for. The vendor is in Poland but I believe the service comes from Dubai. The monthly fee is $1,750 and the speed is about as fast as DSL. It's quite costly but with thirty five users, the monthly fee per person will be little more than home internet. I started a bank account to be used to collect the fees and pay the monthly bill. Before we left home, I collected enough for the purchase of the satellite system, a router, and the first month's service. Fortunately, I have a die-hard military supporter that pledged to donate $50 per month to help offset the exorbitant monthly fees. Thanks Ival!

We spent the rest of the day looking around the FOB. There are buses that continually travel a route on the FOB and make several stops along the way. We hopped on a bus and rode down to the chow tent for lunch. I wasn't disappointed. The food is good. Coming out of the chow tent, we saw the MUF we were told about and stepped in for a quick look. As advertised, it has all the things we were told and more. They have a large movie area and full gym. After leaving the MUF, we checked out the laundry facility and then headed back to the CP. The Commander didn't have anything further for us today so I hung out with their pilots in the CP and spent some time studying the tactical map and learning the AO.

30 June 2007

We got up at 0700 this morning and made a pot of coffee before going to breakfast. While sipping that first cup of Folgers out on the landing in front of our CHU, I saw a departing soldier move a homemade wooden desk out of his CHU and leave it for the taking. With swift action, like an eagle on a wounded rabbit, I snatched it up quickly.

After securing my newly acquired desk in the CHU, I went to breakfast and then attended briefings for the duration of the day. Once again, the Rules of Engagement were briefed, this time by a Jag Officer. There is a great deal of legal responsibility for every action, even in combat. It's an uneasy feeling. Daily, situations will require us to make split second decisions. A wrong decision to pull the trigger could potentially result in a legal action against me. However, if I hesitate and do not fire, it could also be my last decision or cost a life on the ground.

Our Commander published a flight schedule that had our teams mixed 50/50, one of our pilots with one of theirs. For the next ten to fourteen days, we would share cockpits with the departing unit to learn the AO. This will lead up to the official Transfer of Authority, when our squadron will pick up the torch.

The MUF hosted a dodge-ball tournament tonight so one of our pilots quickly assembled a Blackdeath team to participate. We showed up and took names. I'm not sure if winning was the best idea but we did make ourselves known after just over 24 hours on the FOB. Tonight I'm staying up late to adjust my work schedule for the night shift.

1 July 2007

I slept in as much as possible to help shift my sleep schedule for night missions. After eating lunch I linked up with the pilot I was paired with and we went to the TOC for our mission briefing prior to launch. In our intel brief, we heard that one of our crews in Mosul had received small arms fire the previous day. This is a regular occurrence for the Mosul mission.

Our mission today is to conduct local area orientations or LAOs and recon the Syrian border. After receiving our intel brief, we hoofed back to our CP and conducted a team briefing before launching. I had done this before and wasn't nervous, but I knew this was not the same war I left in 2004. I'm in a hostile area and the enemy is much more organized now. I've got my head on a swivel and I'm bringing my A game. Let's do this!

We fired up the birds and repositioned to the FARP for fuel and ammo. We signalled for the pad chief to give us three additional rockets and they quickly loaded us up. We departed the FARP and flew to a test fire range only seven minutes north of the FOB where we each fired three rockets. We squeezed off a few rounds of .50 cal to test the gun and all worked well. We began the LAO right over the city of Tal-Afar and then headed northwest to the Syrian border. We paralleled the border southwest to the western tip of Mount Sinjar. From there, we made a fuel stop at a small jump FARP about 45 miles west of our FOB. We talked about each town, village, and religious / political sect in the area. Before returning to the FOB, we flew to the Peak of Sinjar Mountain to look at a small outpost we had positioned there. The outpost sat at an altitude of 3000 feet. The north side of the outpost was a shear drop and somewhat eerie to fly over. We quickly went from 10 feet above the ground to 3000 feet in only a few seconds. As we descended out of 3000 feet, we called trail and said, "We're Mike Carnage and RTB." Trail replied, "Roger, RTB." This meant, we're mission complete and returning to base. The heat alone is enough to wear you down. I was tired by the end of the flight.

Our AO stretches 70 miles in any given direction. The amount of terrain we are assigned to cover is much larger than I expected. After landing, I peeled my body armor off and my flight suit was soaked in sweat. As the heat dried my flight suit, a clear pattern of white lines formed the outline of where my body armor had been. My body is losing so much salt that I can see it as it dries on my clothes and feel it on my skin. I'm drinking as much water as I can to stay hydrated and I add salt to replenish what my body is sweating away.

After post-flighting the birds, we put them to bed before going to the TOC to debrief our mission with the S2 officer. "Putting the aircraft to bed," means completing the logbook, tying the blades down, putting covers on, and other various items. It felt good to complete the mission and the debriefing; now it's chow time. The chow tent is always a welcome sight after a long flight. After dinner, we went back to the CP to look at the schedule and talk about tomorrow.

I was so exhausted and couldn't wait to get a shower. Sweating profusely all day in the desert heat leaves a nasty, gritty feeling on your skin. Each CHU area has shower and toilet trailers. It's crowded now but should calm down when the other unit leaves. When my head hit the pillow, I was out within seconds.

2 July 2007

I didn't wake up until 1300 today. After some stretching, I put on my flight suit and caught the Tal-Afar metro bus to the chow tent. My breakfast was the lunch meal, but it didn't disappoint. I'm enjoying the food. On the way back to the CP, I stopped at the MUF to use the internet and email my wife. Soon we'll have our own system and won't need to wait in line. It's a pain but well worth the wait, even to send a short message home.

I was scheduled for a split flight today, one mission before dark and one after dark. During the day we practiced throwing smoke grenades from the cockpit and made it interesting by placing a small wager on who could get the grenade closest to the target. The loser would buy free non-alcoholic beer at the chow tent. We always find a way to make things interesting and pass the time. I enjoyed the flight, mostly since I walked away with the victory. We carried four to six smoke grenades in the cockpit and hung them from the door jams by their spoon. The aircraft isn't designed to carry grenades and they were too large to attach to our vests so we had to improvise.

During our night mission, we flew air support for a ground team that conducted a raid on a house in a small village not far from the FOB. During the raid, two men jumped in a truck and took off across the desert. People that flee the mission area are referred to as "squirters." These two squirters were using the wadis, or ditch system, to get away from our ground team. Our job was to round 'em up.

We were the lead aircraft and I was in the left seat. I was on the radio with the ground team and walking them in on the truck. I quickly unstrapped my M-4 and armed the laser. I'll attempt to stop them using a laser they can see. If that doesn't work, I'll fire a burst of warning shots across the hood of the vehicle. My right seater positioned the aircraft to my advantage and I activated the laser, illuminating the hood of the truck. After waving it across the hood a few times, the truck came to a stop. We maintained an orbit around the vehicle and I made the laser dance on the hood of their truck. Our ground team arrived only a few minutes later. They were quickly taken into custody. Chalk another one up for the good guys. I felt accomplished.

We RTB'd just in time to catch some midnight chow. After eating, we debriefed our mission with the intel team and headed to the shower trailers. I was once again mission complete and asleep as soon as my head hit the pillow.

3 July 2007

At 1000 I bounced out of bed refreshed and feeling great. I had a hot cup of coffee in the CHU and started the day at a leisurely pace. A short time later I capitalized on the bet I had won during lunch and drank a cold Coors NA beer. Ah, it was great.

Tonight's mission took us deep into the hours of darkness and north to the Syrian border. We spent 4.3 hours buzzing around, looking for anything suspicious. It was uneventful. The highlight of the night was midnight chow. It's always nice to finish my shift with a good meal in air conditioning.

4 July 2007

Today is Independence Day, the day we annually celebrate our freedom and honor those who paid the ultimate price protecting those freedoms. We just received word that we lost one of our own from Ace Troop. Their Scout Weapons Team (SWT) was flying in support of a ground convoy and one of the aircraft hit a wire, which in turn, separated the tail-boom from the aircraft. As trained, the pilot quickly initiated an autorotation. Although the maneuver was completed as trained, the impact with the ground was significant. The aircraft landed upright. When the dust settled, the result was one pilot KIA and the other critical.

The fallen pilot has a wife and two daughters back home. His family profile was a mirror image of mine. The news hit me hard and we all choked back the tears as the XO told us what had happened. My CHU buddy, Mike, was very close friends with him and has been selected to escort this hero back home. Mike and I went back to our CHU and I assisted him in packing his gear for the trip. Mike was very quiet and noticeably upset as any friend and brother would be.

We lost a great pilot, friend, and hero today. I offered my support in prayer as Mike walked away to board the Blackhawk enroute to Mosul. This scenario resonates in the back of all our minds. Will I be next? It seems like a dice game, but only God knows when our time is up. When it's up, that's it. It's counterproductive to worry about it.

My CHU is now empty and to avoid feeling depressed and lonely in my CHU, I didn't return until 0400 hours. I kept myself awake until I was completely exhausted. I wanted to fall asleep as soon as I laid my head down. I prayed for God's protection over our soldiers and aircraft, and left the light on to combat the feeling of loneliness.

 I read a few scriptures my wife had highlighted in my Bible. God must have guided my hands tonight. As many would do, I placed my thumb on the edge of the pages near the middle and opened to a highlighted passage. "Those who live in the shelter of the Most High, will find rest in the shadow of the Almighty." Psalms 91. I'm thankful she spent the time marking these scriptures. God was speaking to her on my behalf, for this precise moment. It gave me peace.

5 July 2007

It's been but a day since the accident and I'm still in shock. I prayed for his family, friends, and the morale and safety of our unit. I don't understand many things in life, and death. He was a great person, father, and husband. The similarities between his family and mine are sobering. This could happen to any of us. Our Squadron Commander halted operations for a 24 hour period, to allow us to re-cage our minds and get back in the fight. All links and means of communication to the civilian world have been blacked out for a full 24 hours. The blackout period is to allow the family members to be properly notified of their loss in the most appropriate manner. Today has been a very quiet day for 4-6 Air Cav. I stayed up late again tonight, in an effort to go to sleep quickly.

6 July 2007

The communications blackout is over and I was able to call home. My wife wasn't a close friend of his wife but we all know one another. I'm sure she has the burden of thoughts similar to mine. I must have said "I love you" to the girls twenty times. They probably thought I was crazy.

I received mail today. It was the highlight of my day. The girls sent pictures for me to hang in my CHU. Now I can see them daily as I'm waking up. I love you girls with all my heart.

7 July 2007

I'm trying to get adjusted to a new sleep / work schedule, and it hasn't been easy. The last two days have been very stressful. I was able to chat with the girls today, except for Maddy, who is spending the night with a friend. I have to get my mind back on track and off of hopeless possibilities. Many times, the mental war is bigger than the physical war. We've been gone a month now. A month doesn't seem that long since we have fourteen more to go. I'm already thinking about R&R and how I'll enjoy my days with the family.

8 July 2007

I'm back on a day schedule with a very early show time. It seems twice as early, having been on nights. I propped my eyelids open with toothpicks and made a strong cup of coffee to slap me in the face. Shifting sleep schedules sucks for about three days.

Our new and up-coming internet system has been the talk of the Troop for a while now. It won't be long until 1-17 Cav is gone and we take over the system. I usually link up with Bertie in the evening and have the ability to use the web-cam most days. It's similar to coming home after work, but instead of being in person, she's on my computer. Having a link to home and the outside world is a morale booster that is immeasurable.

9 July 2007

We're getting a bit more comfortable with our mission area. We took small arms fire during our mission today. As we heard the burst of rounds ring out, we called our trail ship to cover us, "Taking fire, breaking right!" Our wing rolled in and suppressed with a burst of .50 cal. We quickly assessed the aircraft for damage as we flew out of their weapons range. There were no apparent issues with the aircraft so we continued and tried to get positive ID on the source, but were unable to. It would be easier if they would stand up and fight instead of cowardly firing a few rounds and then melding back into the population. My heart rate jumped up after the incident and my senses were at their peak for the rest of the flight. Upon landing at the FOB, we carefully inspected the aircraft for damage and fortunately found none. I thanked God for protecting us.

10 July 2007

It's another hot day in the desert with temps reaching 115°. I must have sweated off five pounds of water weight. We spent several hours flying air support for a large convoy moving through our AO. After flying over the convoy for two hours, we made our way to the FARP to refuel. After arrival at the FARP, I gave the pad chief the number of gallons we needed. As the rotors turned overhead, they began pumping fuel into our bird. The FARP crew over-filled our aircraft with fuel, which is irritating because I had to get out of the bird and off-load two rockets so we wouldn't exceed our max gross weight. While I was unloading two of the rockets, I noticed that they had also failed to put our fuel cap on. Now I'm thankful they over-filled our fuel or I may have not noticed the dangling fuel cap on the side of the aircraft. Who knows what would have happened if I hadn't seen it. Thank the Lord, today my number wasn't up. After a very long flight, we were able to safely escort the convoy through our AO and RTB. I chatted with my girls and am thankful to be alive.

11 July 2007

We only have a few days until we take over the satellite internet so I spent the majority of my day off preparing the equipment and planning the installation. With 35-40 users on the system, it will take some time to build and run all the cables to the CHUs.

I spent time on the internet with the girls and Maddy was sure to tell me what she wanted for her birthday. She had carefully thought it out and decided she wanted to go to a water park. Bless her heart; I'm sure she'll change her mind tomorrow. I didn't care. I was happy to see her face on webcam.

15 July 2007

Knowing how important the internet system is to our troop, the Commander pulled me off the flight schedule for a few days to allow me enough time to install the hardware for our troop. I had a good bit of help during the process. Everyone is eager to get it operational.

The system is now up and running. We have internet in our CHUs and the troops are happy. As expensive as this thing is, it's well worth it. Now that our system is up, we have several people trying to gain access to it. With limited bandwidth, we have to be selective on the number of users we have.

25 July 2007

I haven't written much lately. With the convenience of internet in my CHU, it's the best social outlet and gets my mind away from this place quickly. I chat with the girls and dream about R&R.

Most of our missions have been fairly routine and without incident.

We were recently out on mission and were called to a convoy for immediate air support. The convoy was moving through a small town and some very unfriendly people began to gather around them. The convoy commander had that familiar and uneasy feeling that something bad was about to go down. As we arrived, we made a couple of very low and intimidating passes over the groups of people. They quickly disbursed and nothing became of it. The insurgents are hesitant to attack any convoy with us overhead. Ask any convoy commander who they want for air support and they'll tell you, "The Kiowas." We stayed with the convoy to the boundary of our AO. Nobody messes with our boys. You mess with them and you'll have to deal with us. We're looking for a fight.

During our intel briefings, we hear about the enemy activity in Mosul. It sounds like Ace Troop has had several small arms engagements. Our Commander said we will soon be moving into the Mosul area for missions. It hasn't happened yet but it's sure to come anytime.

There's a lot of talk about when we may possibly pull out of Iraq. Many believe it will be around the presidential election back home. Promotion rates in the military are through the roof because so many people are getting out. The Army is having difficulty keeping soldiers in so they're offering big bonuses to stay. Two of our soldiers reenlisted and received $27,000 bonuses. I'll be up for CW3 soon and will have to make a decision to either stay and accept promotion or get out.

In my down time, I've been working on a video to send back home to the girls. It's almost complete. The girls should enjoy it.

27 July 2007

Today's mission placed us near the south eastern border of our AO and south of FOB Q-West. We were tasked to recon several objective areas, roads, bridges, and supply routes. We covered a lot of territory today and fortunately it was uneventful. We made a fuel stop at the FARP on FOB Q-West and departed to the northwest. On the return leg of the flight, we conducted a route recon of one of the main supply routes leading up to Tal-Afar. The return flight would take about thirty minutes and we should be hitting the chow tent not long after landing, or so we thought.

Approximately ten minutes after take-off, our TAC sent a text message to the aircraft and advised us that a dust storm was inbound from the northwest and was moving in rapidly. We were on converging courses with the massive dust storm and it was a race to see who would get to the FOB first. We could see the top of the storm in the distance and had pulled as much torque as the birds could muster. An intimidating wall of dirt was moving our way at roughly 45 mph. It was only a couple minutes later that we realized the storm had already overtaken our FOB. The lead ship made the call to deviate course to the north in an attempt to get around the frontal wall of the storm. In only minutes, the front of the storm was just out our left door and the mountain range that was five kilometers away, had disappeared in the dust. The storm was far too wide-spread to navigate around so the only other option we had was to change course and fly directly away from it. We made a hard right turn to the northeast and set a course for Mosul.

Mosul was approximately forty-five kilometers to our twelve o'clock and getting closer every second. Flying to the limits of the aircraft, we quickly put some distance between us and the storm. It only took us about twenty minutes to get to Mosul and land. We quickly shut the birds down and went into the Mosul TOC to check weather. After the dust storm passes, it would take at least a couple hours for the air to clear enough to fly again. Knowing we would be down for a while, we headed for some chow and cold drinks.

When a few hours had passed, the air cleared enough that we were able to fly back to FOB Sykes. We ended the day with 4.3 flight hours and sore backs. After debriefing the mission with our intel team, I headed to my CHU for down time. I linked up with the girls on the web and chatted for a while. Having internet on demand, in my CHU, is a priceless commodity.

While lying in my bunk winding down, I reflected on the war and how it differs from World War II or Vietnam. Soldiers from previous wars had it much worse than we have it. They were lucky if letters from home found them on the battlefield. They lost many more soldiers than we have and the conditions were harsh compared to ours. I hope our mission here is making a difference for the people of Iraq. Without saying too much, the information gathered in Iraq has protected Americans back home and debunked planned attacks on U.S. soil. In many respects, this is a war for information gathering. We're fighting a cowardly enemy that hides behind women and children to protect themselves. The insurgency is in short supply of suicide bombers and often times recruits men by terrorizing their families and threatening to kill them all if they refuse to cooperate. I pray I don't kill any innocent people while I'm here.

In just the short time we've been here, I've had to make a difficult decision of whether or not to pull the trigger on someone. The current ROE, or Rules of Engagement, state that if someone is found damaging the country's infrastructure or vital systems, such as the oil pipeline, they can be engaged and killed. We happened across a man tapping into an oil pipeline and siphoning oil for his own use. Although legal to engage, I passed on the opportunity and called it in to our ground patrol team to check it out. I later learned that with many people out of work and going hungry, they are attempting to provide for their families by stealing government oil. In my eyes, it's a crime hardly worth killing for.

9 August 2007

It's been several days since I've put pen to paper. Madison's birthday is coming up so I bought her a jewelry box and shipped it home. She just received it and was happy to get it. Every kid likes early birthday presents. She's been bugging her mom non-stop about her birthday. She's as excited as they get when it comes to birthdays.

I had the pleasure of flying with our Squadron Commander today. I enjoyed the flight. We talked about hog hunting in Louisiana while we skimmed the earth at 100 knots, looking for bad guys and suspicious activity. I've never been hog huntin' so he shared his stories, and the flight was gone before I knew it. There have been several more aircraft shot at since my last writing. Thankfully, none of the pilots have been hit. Our aircraft have sustained some damage, but those are much easier to fix than pilots.

14 August 2007

While my daughter is home celebrating another year of life on her birthday, suicide bombers attacked two small towns west of our FOB, Kahtaniya and Jazeera. The attack consisted of four suicide bombers in separate vehicles that drove into the middle of these two towns and detonated their VBIEDs. The attacks were the result of a conflict between the Yazidi and Sunni Muslim people. The death toll was tremendous. The initial body count was reported as 300-400 dead.

We were just finishing our mission that evening when we received the call. We were tasked to provide immediate air support to both towns and send a battle damage assessment, or BDA, to the TAC. The plumes of smoke from the blasts were easily visible from miles away. Upon arrival, the scene was extremely chaotic. The blasts levelled houses in both towns over a radius of several blocks. There were bodies lying in the streets and people digging through rubble in an attempt to save their loved ones. Vehicles were pouring out of town, trying to get their wounded to the closest hospitals, while ambulances sped toward the scene. The ill equipped band aid stations were quickly overwhelmed and began turning people away. Approximately an hour to an hour and a half after the blast, locals began showing up at the gate of FARP NIMR, west of FOB Sykes. I heard Captain Young calling for medevac support. The TAC was slow to respond and frantic. I could hear the desperation in his voice as he said, "I have people dying on my steps!" Our medevac birds are typically kept in reserve for coalition troops, but today our command group made an exception to policy and launched the Blackhawks. Captain Young and his soldiers were in no way equipped for this level of mass casualty response. There were approximately forty people laid out across the blood soaked helipads. With the refuel pads blocked, our closest fuel point was FOB Sykes, a twenty minute flight away. We were near the end of our mission window and low on fuel so we conducted a battle hand-over with our Troop Commander, Blackdeath 6, and returned to FOB Sykes. Below is the account from the ground Commander at COP/FARP NIMR, Captain Joe Young.

My name is CPT Young, Commander of Ghostrider Troop (Ghostrider 6), 4-6 Air Cavalry Squadron. This is my account, of a hot August night, when heroes shined bright. Elements of both Ghostrider Troop 4-6 ACS and B Troop 1/9 Cavalry were co-located at FARP NIMR this day. This was a combat outpost established just below the Sinjar Mountains and near the Syrian border. We had just finished barbequing some steaks on an old grill my soldiers found. It was a welcomed break from the routine diet of MREs We washed them down with some Rip-Its. Rip-Its are a foreign brand of energy drink. We were moving into our initial security posture for the night when at approximately 1930 hours, we heard four large explosions in the distance. I felt the shockwave in the ground and saw multiple plumes of smoke in the distance. A wave of tension moved through by body and I knew we were in for a very long night. This night changed our lives forever.

The tower guards reported the blasts to our higher command on FOB Sykes. Based on the events, we established a 100% security and defensive posture Based on the events, we assumed a defensive posture. SSG Sonnenberg, SGT Garay, and I mounted up in a gun truck and positioned ourselves at the front gate of the compound. We monitored several reports over the radio, shortly after the blasts. The airwaves were filled with chaos and reports of mass destruction in two nearby towns which included reports of 100 enemy insurgents within three kilometers of our location.

It was said that VBIEDs were the cause of the destruction and that the two helpless towns suffered massive casualties in a matter of minutes. 1/9 Cavalry didn't have the assets required to evacuate the wounded to a higher level of care. A report came over the radio that there was still one VBIED at large, but they had no physical description of it. Flying above us was Blackdeath 6, Commander of the OH58D Troop based at FOB Sykes, and his Scout Weapons Team. They were calling in reports to our higher command and keeping us abreast of the situation from the air.

Locals poured from the affected towns and began making their way toward our COP. Blackdeath 6 estimated that nearly 100 people were coming our way and were presently only three or four kilometers away. We fortified the perimeter security with all remaining personnel and gun trucks. The outermost perimeter, located about 400-500 meters away and offering security for the COP, is usually manned by Iraqi Army soldiers (IA). This area was mainly built up dirt mounds with a few guard towers. Only the actual checkpoint along the paved road was fortified with concrete walls and barriers. Most of the IA guards had left the area to assist with rescue efforts in the two towns. I ordered the guards to fall back one level and establish security at the first set of HESCO barriers. I ordered PFC Black, SPC Anderson, and PV2 Cook to man the .50 cal and (2) 240Bs in guard tower and prepare for an imminent attack.

We remained ready in this posture for 30-45 minutes before the first group of little minivans arrived. We could see a few blue and red strobe lights flashing in the sky like police vehicles, but we could not see what the vehicles contained. We were truly concerned about the remaining VBIED reported in the area.

An IA truck drove up next to the large door at the front gate and an Iraqi Soldier jumped out and began yelling something. Communications were extremely difficult because we had no interpreters on site at the time. I finally understood that he was requesting helicopter support to evacuate their wounded. We tried to explain that there were no helicopters at the COP. I was concerned that requesting a MEDEVAC aircraft would feed into the plan of the insurgency and ultimately be shot down. I'm fairly certain insurgents were plotting our movement as we drove to and setup operations in the COP. The Iraqi soldier hopped back into his truck and left the compound.

Within five minutes we were bombarded with multiple ambulances and civilian vehicles. Once again, we could hear someone yelling on the far side of our door. We knew multiple vehicles were there, but were still concerned about the VBIED that was yet to be accounted for. However, this time we could hear the faint cries of small children in the back ground. The scene was getting chaotic.

The gate was braced by a gun truck. I moved everyone but myself and a couple others away from the gate area and elected to take a chance in an attempt to help them. I pulled far side security with my M-16 and SGT Garay had the near side with his rifle. Exercising extreme caution, SSG Sonnenberg and I stepped one meter out of the COP to assess the situation. Under the glow of the minivan's dome light, we captured first sight of the gruesome after math. The night sky in the background was littered with explosions. Many victims lay tattered and bloody in the mini-vans, but what caught my eye was a father holding his dying son in his arms. The young boy had extreme head trauma and blood was pouring from his wounds. Without hesitation, the three of us immediately opened our gates to the wounded. We immediately began administering aid and tried to triage the best we could in an attempt to stabilize them until we could get additional help.

We had a blood bath on our hands no doubt, but we still had the responsibility of exercising as much diligence as possible to keep our team safe. SGT Boldoe and SGT Smith took the initiative and came forward, searching every vehicle from top to bottom before each was allowed to enter the COP. Even though they confiscated many weapons coming in, it later became evident that the locals were concerned only for their own welfare and showed no resistance to surrendering their weapons. The last VBIED reported in our area had still not been found and this concerned us all.

Due to the number of casualties that arrived in the first wave, SSG Sonnenberg, along with SGTs Garay and Richard, quickly expanded the CCP to accommodate the injured. All three were crawling through blood covered vehicles to triage patients by order of priority. I initially rendered treatment as well, but soon realized that many were in need of evacuation to a higher level of care. I made the initial request for MEDEVAC at approximately 2115. By this time, the number of wounded was over twenty and rising. I initially received hesitation from the distant end of my headquarters because the injured were Iraqis and I was told to take them to a local hospital. I informed them that the hospital was blown up in the blasts and doctors there were already overwhelmed with treating those they could. I was told to wait on MEDEVAC due to political implications and I lost my composure and asked if he, as the Squadron S3, was telling me to let these people die in the arms of my soldiers.

At that point, Blackdeath 6 came on the radio and painted a picture that the S3 couldn't seem to comprehend from my previous reports. This was no small event. This was a MASCAL! I threw the hand mike down after repeating the nine line a couple times. I was physically shaking as I moved out to check on both the soldiers that were treating everyone and the soldiers who were pulling security.

I grabbed SPC Krockmal and we began documenting each of their injuries while treating them. At 2130, a second request was placed for MEDEVAC support. Only four of our soldiers had received the Combat Life Savers (CLS) course prior to deploying to combat, so our experience was limited. As the CLS trained soldiers treated the wounded, other soldiers watched and quickly learned the critical steps so they could began treating casualties too. We also only had 4 CLS bags to use medical supplies from. The supplies only lasted through the first 5-10 people. When those supplies expired, we used t-shirts, BDU coats, bed mattresses, 550 cord, duct tape, toilet paper, and whatever we could find to make it work.

Casualties continued to pour in. We treated patients with exposed brain matter, shrapnel wounds in all regions of the body, splinted broken limbs, dressed open chest wounds, applied tourniquets, treated bleeding intestines, removed AK-47 rounds, and administered IVs. SGT Garay scrounged up *some twine and used it to build an improvised gravity fed IV device to keep fluid running to patients.*

PFC Krockmal and I began prioritizing the casualties for the MEDEVAC bird, and then loaded them onto the back of an LMTV. Many of these patients were children ages 3-4 years old. Due to the limited number of litters, we had to improvise and use tables, plywood, and mattresses to move the wounded. In some cases, we moved 2-3 children on one litter to maximize our efforts. After loading the wounded into the LMTV, PFC Bayan and PFC Allenza drove them to the landing zone. The landing zone was marked with infrared chem-lights.

I called the MEDEVAC aircraft and gave them an LZ briefing. The birds landed shortly after my radio call and we began briefing them on the order of priority. Every time the bird would land, it would blow dust over everyone and cause chills in the little children. We grabbed what we could to cover them up. They loaded our wounded and departed. We still had reports of enemy dismounts in the area, as well as the remaining VBIED, so we had to man the .50 call on the LMTV and my gun truck to escort the LMTV back and forth from the triage area. This assured security for the patients and for the incoming aircraft should we be attacked.

When we returned to the triage area, I was briefed that we were running low on medical supplies. We continued treating people and a second wave of wounded were evacuated. During the third iteration of loading wounded into the helos, two mortar rounds impacted just northeast of the aircraft, creating a plume of dirt that engulfed the lighting from the Special Forces compound adjacent to ours. I saw the crew chief from the lead ship move quickly to the pilot to see if he was hit. Fortunately, neither the pilot nor the aircraft had been hit. There was a large fuel truck not far from our location, thankfully it too was unharmed. I remember hearing the impacts and feeling the shockwave as it pushed me into the side of my gun truck while I stood there with a hand mike to my ear, talking to the aircraft and our HQ back at FOB Sykes. I remember looking at SGT Morris and yelling at him, "Did you see that!" I don't think he heard me over the rotor wash. I looked up into the night sky and asked God what else he was going to throw at me and my men. On the fourth wave, the Ghostrider Troops along with B Troop 1/9 Cavalry, loaded the remaining casualties into the MEDEVAC birds and sent them one last time into the black sky.

Many more ambulances came to our location. Out of supplies and options, I had to make the difficult decision to turn them away. By this time we had an interpreter so I asked him to redirect them to a hospital in Sinjar, a town not far from our location. Several people died at the tips of our fingers. They were moved aside so we could care for others. The bodies of the deceased were taken from our location by the Iraqi Army. It was approximately 0230 hours before the dust settled and we could begin processing the events that had just occurred.

The Kiowa Scout Weapons Team provided aerial security for a short window of time. I gathered everyone together and they stood almost mindless, mentally depleted, and covered in blood from head to toe. Some were missing ACU tops and many wore boots that were soaked with blood. I sent everyone to shower off as I pulled security. SSG Sonnenburg tried to clean up the triage area because of the sights and smell of death and carnage that resonated from it.

Once everyone had showered, I took them down stairs and inside our TOC to conduct an after action review (AAR) with everyone. This was an effort to capture all that had occurred, but also to get an initial assessment of everyone's mental state. I told everyone that this would be something they would remember for the rest of their lives, and if anyone had any issues tomorrow or next year, talk to someone about it. I told them that I was extremely proud of them and that they should be proud of themselves. I recognized PFC Smith specifically as the hero of the night. Of everyone, he was the most saturated in blood from body parts. This average performer cranked out a 150% effort during that gruelling twelve hours and impressed us all with his humanity and compassion. I asked him where that came from and he said, "It was the right thing to do." I'll always remember that about PFC Smith.

We picked up our security rotation for what was left of the morning hours. I pulled the first shift as everyone cleaned themselves up, and then rotated out to shower off. I think I faded off once or twice that night. I couldn't stop shaking. I began to fear for my own children's lives. When the sun rose, SSG Sonnenburg came to me as we were burning the bloody uniforms and said the soldiers were already beginning to show signs of PTSD. I contacted my 1SG and told her to prepare the next team to replace us. We made a request for the chaplain to visit the COP. By the end of the next day, I had scheduled a relief team and also told the mental health doctor at FOB Sykes to expect 19 personnel for evaluation.

Our experience was not the kind of war I envisioned. Seeing the dying children and hearing their screams echo through the darkness still infects my dreams and the dreams of those soldiers who stood with me that night. I stay in touch with many of them today.

At least half have admittedly had struggles with PTSD. One member tried to commit suicide twice and thankfully, failed. The last time I spoke to him, he was getting the help he needed. I have nightmares from the screams, faces, and injuries. I will never forget the face of the poor man that brought me his dying son. The smell was horrifying. The events that night formed an unbreakable and life-long bond within our team.

I submitted all the soldiers for the Army Commendation Medal with Valor, for their actions. All were downgraded to the standard ARCOM. I feel they deserve much more and I'm sure the survivors would agree.

COP/FARP NIMR Team

When the dust settled and the final count was in, there were nearly 800 dead and over 1500 wounded. It's been labelled the second deadliest terror attack in history, behind the 9/11 attack in New York.

Chapter 13

Fallen Angel

15 August 2007

After breakfast my team met at the TAC for our mission briefing. The Iraqi Ministers of Security and Health flew to FOB Sykes today to deliver large sums of cash to the two affected towns. When they arrived, they were toting two large cases with $300,000 in each. A team of our Blackhawks was tasked with flying the two officials into the towns, and we were tasked with providing security throughout the movement until mission completion. If word gets out that we're delivering that much cash, this mission could get bloody in a hurry.

Like clockwork, we fired up our birds and in minutes were lined up on the taxiway, ready for take-off. A short time later, the Ministers moved from the TAC and boarded the Blackhawks with $600k in tow. Immediately after take-off, we moved into formation. The Hawks were flying lead and our Scout Weapons Team was loosely trailing. We positioned ourselves to over-watch the real estate forward and lateral of the Hawks. Within twenty minutes, we arrived at the first landing zone. Upon arrival, the Hawks landed near the edge of town and dropped their passengers. We surveyed the perimeter for suspicious activity. Now daylight, we were able to get a better assessment of the damage. From the point of detonation, all buildings were leveled within a four block radius. Beyond that, there were buildings damaged up to ten blocks away. We took several pictures for our intelligence group to analyze. This has been documented as the most deadly attack in Iraq to date.

The Ministers weren't on the ground long before they boarded the Hawks. The Hawks quickly departed and made a short hop to the second town and repeated the process. The initial intel coming to us pointed to political tensions between two groups. The hostility and lack of consideration for human life here is mind-boggling. I'll never understand it. After the "packages" were delivered, the Hawks lifted and we escorted them back to FOB Sykes. Once safely on the deck, we topped off our fuel and continued the mission. Much of our day was spent providing aerial security for the two towns, in an effort to prevent follow-on attacks. The day ended uneventful and I'm thankful for it.

21 August 2007

It's been a very busy week and we've flown the blades off these birds. We had several QRF launches across the squadron as a result of ground troops coming under fire. They are quick to launch us because, as soon as we appear overhead, the violence stops. You can't out run a helicopter and we carry the firepower to rapidly eliminate the enemy.

The Iraqi Army has been more proactive and conducting many raids. Our ground troops have been leading and over watching Iraqi troop operations. The ultimate goal is to train the Iraqis to be self-sustaining and courageous enough to stand up to terrorists. In the initial raids, they would kick the door in, but as soon as bullets began to fly, they fled in fear. They are improving, but the progress is slow.

A small group of us began this deployment by growing mustaches and calling ourselves "Stache Team 6." Our Squadron Commander thought it was a great morale builder and supported us by holding an unofficial back-office tabbing ceremony in Kuwait, before moving into Iraq. The SGM was less than happy to see the unauthorized tab on our uniforms and made a big stink about it. Today marks the official disbanding of Stache Team 6. We continue to rebel by secretly wearing our tabs under a pocket flap. It's the small mental victories that keep you going.

22 August 2007

I bow my head today and say prayer for the families and friends of fourteen soldiers killed when one of our Blackhawks crashed while conducting a night vehicle interdiction mission. I knew one of the pilots well and was friends with the other. Both were great men and aviators that heroically served our country. We had two very dedicated crew chiefs on the aircraft that night. Along with the crew, ten ground soldiers were killed doing what they loved. No words are can explain the devastation our unit is feeling. Cory, Josh, Matt and Ricky: you will be missed.

They had just completed their last interdiction for the night when they lifted and began to spin out of control. Based on the preliminary reports, the aircraft impacted the ground at a high speed and there were no survivors.

Although tragic, my daily routine and exposure to death here has desensitized me. We all know this can happen to any of us, and continue to spin the roulette wheel of life. We're here to do a job, so I'll take a deep breath, say a prayer for the families and loved ones, and have faith that God will continue to protect me. We're currently under a communication blackout until the families have all been notified. This is always a tough time for our families back home. They know something bad has happened when a blackout occurs. All they can do is wait and hope that they don't get a knock at their door.

30 August 2007

It's been several days since I've written anything. Our flight schedule has been rigorous. I've flown nearly 100 hours this month. Flying this much causes fatigue issues so we have to watch one another for signs and heed warnings from fellow pilots. We recently had two pilots make mistakes that resulted in a torched engine and complete drive train replacement. Thankfully, neither was injured.

While I've been busy here, my parents have been dealing with health issues back home. My Grandmother recently moved in with my parents and my dad has experienced health issues of his own. He's a surviving cancer patient but has ongoing issues from the treatments.

The changes are stressful for my mom, but she's coping. I need to call home and check on everyone.

17 September 2007

We're now in the period of Ramadan. Ramadan is a period when Muslims fast and deepen their spiritual relationships. We tend to see an elevated number of attacks during this period and are busier than normal.

We had a great mission yesterday. We were supporting a Special Forces ground team while they conducted a recon on an abandoned house in the hills north of the FOB. The house was being used as a rendezvous location for Al-Qaeda and sometimes as a training camp. There was an abandoned vehicle at the location. It was identified to have been used in an attack against coalition forces. At their request, we engaged the vehicle with rockets. There's satisfaction in knowing they'll never be able to use that vehicle as a weapon.

1 October 2007

Our days continue to be mission rich. Our intel group has been spot on and we've rolled up several terrorists operating in our AO. With each terrorist caught, I know I'm protecting the lives of our soldiers here in Iraq and that makes everything worth it for me. Most of the raids are fairly routine. Our ground teams hit a house and the enemy either gives up or jumps in a vehicle and attempts to run. If they make it past our ground team, we'll round 'em up from the air.

Since the beginning of this deployment, I've been smoking cigars on a regular basis. It relaxes me and calms some of the tension. I recently sent a batch of cigars to my uncles in Missouri so they could celebrate the expansion of their auto repair business. They're moving into a larger building and restructuring the business somewhat. Congrats boys!

I also ordered a new computer network switch for our internet system and network. The old switch didn't allow me to properly manage the bandwidth and ensure users were only using their fair share of data.

The supply folks are probably tired of seeing me. I've been over there daily, checking on its arrival. It arrived today. I installed it and its working great. Now it's time for a cigar. Ah!

7 October 2007

The network system is operating much better since the addition of the new equipment. We're still having issues with power outages. Our CHUs are powered by two very large generators and each day, the maintenance men shut one generator down and start the other generator. This process takes about six minutes. The frustrating part is when you're in the middle of something on the internet, such as a chat session back home, and the power fails! The maintenance cycles are unannounced and aren't very timely. I ordered a battery backup unit that is able to run all our equipment for up to twenty minutes without interruptions. It arrived today so I'll get that installed and hopefully boost morale a little more.

We've settled into a routine and the time seems to be flying by. It won't be long and I'll be headed home for R&R. We've been gone from home for five months now. It doesn't seem that long. I'm comfortable in my new environment. It's amazing what you can get used to.

15 October 2007

I've had a couple days off and spent more time talking to the girls back home. Ramadan is now over so the attacks should decrease. The enemy activity in Mosul has been raging. Early this month, our Squadron Commander was nearly hit by an AK-47 round that came up through the cockpit between he and the other pilot. Fortunately, neither was hit. Only two weeks later, he took a round through the helmet that narrowly missed his head. The bullet entered his helmet, cutting through every layer of his helmet liner and skimming the paper thin foam that lay against his head. The bullet missed his skull by 1/8". Looking at the entry and exit holes, I would have bet money that it hit his head. God was protecting him. He said when the bullet hit his helmet, it felt as if someone had hit him in the head with a baseball.

After talking to my wife, I can tell she is a little stressed right now. It's not easy taking care of everything back home without help. Hang in there girl. I'll be home soon.

26 October 2007

It's only October but I'm already anticipating my trip home at Christmas. I've been buying presents for the girls. They wasted no time sending me their Christmas lists. Reading these lists, they must think we're made of money. At the top of Madison's list are a laptop and an iPod. Are you kidding me?! What happened to asking for comic books and dolls? My girls are growing up without me.

5 November 2007

We're now in November, another month closer to my December R&R and Naelyn's birthday. I talked with the girls today and it refilled my heart. It sounds like they're doing well. My sister-in-law, Allison, sent me some goodies from her work team at Kroger. It was very thoughtful of them and I'll definitely enjoy it. Thanks, Allison!

Our Squadron Commander has begun a battle rotation between our Troop and Thug Troop in Mosul. Every week, we are swapping two pilots with them. Our missions differ greatly so the Commander wants us to mix the experience and learn from one another. Captain Rosnick is currently with us and has graced me with a few stories from his Afghanistan deployment last year. He was tasked to work with a ground team and was imbedded deep into the village population, a very vulnerable position to be in. He had some crazy stories, some funny and others horrific. I smoked a cigar and listened intently. After swapping stories, we mounted up and continued making memories in the air.

13 November 2007

I had the day off today and I spent it reading, talking to the girls, and relaxing. The days I'm working don't allow for such things, so I tend not to be too homesick until I get a lot of free time on my hands, much like today. I'm anticipating my R&R since it's just over a month away.

I've been in deep thought of whether or not to stay in the Army. I weigh the pros and cons and lately have been leaning towards getting out. I've been asking for a move to Fort Rucker and the Instructor Pilot course but our career manager won't tell me anything. Figures, they tend to ignore you until you're about to get out or they need something from you. I'm in a very rewarding job and hate to give it up. The transition out is sure to bring me some depression. I can hardly imagine anything else being quite as rewarding.

24 November 2007

The chow hall prepared a large Thanksgiving meal for us two days ago. It was quite impressive and the turkey was good. Good food is a must. A great meal after a mission is like finding an oasis in the desert. It's a place to share stories and even a near beer now and then.

Mom wrote me a letter and filled me in on all the news back home. They're in a very stressful situation right now with my dad and grandmother's health declining. With my dad's health issues at hand, his business is in trouble as well.

2 December 2007

I'm flying nearly every day. Our recon efforts are paying off. We're finding more homemade explosives, better known as HME. The trick with HME is that it has to sit in the sun and dry. In large quantities it's easy to spot in the open, so they attempt to put it down in ditches or wadis, hoping we won't see it. From the air, it's fairly easy to spot. Every pound we take out of their hands could be another life saved.

I was off the flight schedule this afternoon so I watched a movie in my CHU to escape my world for a while. I watched 'A River Runs Through It' and it reminded me of the great times I spent with my dad on the water, making memories. The movie displays how deep a fathers love can run for his son and how it can be never ending.

This struck an emotion in me and I sent my dad an email thanking him for those times. It took me over thirty minutes to type it. I was having troubling reading through the tears rolling down my face. My situation has given me a new appreciation for everything good in my life.

8 December 2007

Thug Troop, in Mosul, has seen a lot of action in the past couple months. Like clockwork, terrorists try to bury IEDs in the roads nightly. As fast as they put them in, Thug Troop is hunting them down to eliminate them, keeping our ground teams and locals safe.

Captain Rosnick told me they were flying over one of the test fire ranges and saw a man lying on the ground. They flew in to check on him and he had picked up some unexploded ordnance or UXO. When the man picked it up, it detonated and blew his hand off. Additionally, he sustained a severe abdominal wound. Captain Rosnick immediately rendered first aid and called for Medevac support. The Medevac team was launched and the man's life was saved due to their actions that day.

30 January 2008

I caught a flight out of Iraq and left for R&R on 26 December. It took me a couple days to get home. I flew into the Seattle airport and deplaned as quickly as I could to find my girls. They were standing there waiting for me as I came through security. It was an emotional moment as we hugged and shared a few tears.

After a few days with the girls, Roberta and I went to Seattle and spent the night by ourselves. We spent the day leisurely exploring the city and the night in a hotel room over-looking Elliot Bay. We had dinner in the elegant hotel restaurant on the main floor. Seated by the window, we sipped a glass of wine while we watched the reflection of the city lights dance on the water of Elliot Bay. Our fifth story room was equipped with a fireplace and a small balcony. After dark, we walked out on the balcony and had a beautiful view of the city skyline. What a difference this was from my life the week before.

I thoroughly enjoyed my time at home. Waking up with my wife next to me, sharing coffee time, and walking by the lake in our serene neighborhood all meant more than it ever had. Everyone needs time away from the stresses of combat.

11 Feb 2008

I made it back to FOB Sykes safely and just in time for our first rocket attack. The attack consisted of a single rocket that detonated near the chow hall. Enemy activity in our AO has been on the rise. We've intercepted much of the smuggling activity which has in turn pissed them off. Our intel group said there has been some talk of an overwhelming force of 300 plus that are planning to attack our FOB. It's probably all talk but we have to take it serious.

I spoke to my dad recently and he said my grandmother isn't doing much. At nearly ninety-two, I shouldn't be surprised. I pray that both he and my grandmother get better. Many of the guys have ordered RC cars to play with and keep their minds occupied during down time. It looks pretty entertaining so I ordered one too. They have a small track built in our area. This should be fun.

17 Feb 2008

We received word from Thug Troop in Mosul that one of their pilots was shot in the leg during their mission. Their aircraft was hit by five AK rounds, one of which struck Captain Sickler in the calf. CW2 Russell flew the aircraft directly to the emergency Cache helipad and immediately applied a tourniquet to it.

Events like these snap my mind back to the reality of what can happen at any minute. Strangely, it still feels like going to work anywhere else. It's scary what you can get used to. Lt. Sickler will leave the fight and be shipped off to Germany to a higher level of medical care. The great thing is that he's going to be okay.

We've had multiple rocket attacks on the FOB recently. Funny enough, I was only aware of the one I previously wrote about. Apparently, the others were so far off target that it wasn't immediately evident that they were attacks.

11 Mar 2008

We had an interesting mission yesterday. We were flying air support for one of our Mad Dog ground teams looking for a weapons cache of rockets in the hills overlooking Tal-Afar. It was said that an Al Qaeda group had stashed several rockets in these hills to prepare for future attacks. We managed to find the cache and along with it, an IED placed on the walk trail up to the cache. The IED was packed with enough explosives to destroy a vehicle. It was rigged to kill personnel approaching the cache. Mad Dog requested we destroy it with a hellfire missile since the EOD team wasn't available and it would have taken several hours to get them up there. We received clearance to fire a missile and destroy it. I love my job!

The temperature is warming up. It's been up in the eighties and it's only March. I received word from home a few days ago that my dad has cancer again. He has several tumors in his bladder. The cancer had progressed rather quickly so they did immediate surgery and removed it. It was invasive so they removed his entire bladder. As I write this, they're doing a bone scan and MRI to determine the extent of the damage. I pray they find nothing additional.

I'm scheduled for a rotation to Mosul in a few days. It would be nice if I could get some closure on this, and know my dad will be ok before I go. I need to have my mind right in Mosul. They've sustained many attacks on the FOB in Mosul and experience more surface to air fire on the aircraft than we do in our AO.

12 Mar 2008

I drug myself out of bed around 0500. Our team linked up and went to breakfast, then to the TOC to get our mission briefing, followed by the flight line to pre-flight the birds, and then back to the CP to conduct our team briefing. We were tasked with conducting a route recon just after day break in the north-eastern part of the AO. It was a beautiful morning and the sun was just beginning to rise as we departed the FOB.

We flew toward Mosul and neared the edge of the AO boundary looking for suspicious objects or activity along the route. We widened our flight path and began scanning the farmland in the area when my co-pilot spotted something that was slightly out of place. We circled back to the area and noticed the dirt was slightly discolored in one area from the surrounding dirt. The area in question was a large bare dirt field of approximately fifteen acres. The border of the field consisted of a five to six foot dirt berm. After two passes and a bit of discussion, I said, "we need a better perspective." We climbed up to about 400-500 feet and it made it much clearer. There were several sections of dirt that were a different color. As we discussed the oddity of it, we determined that we should call for a ground element to come and investigate.

There was a small village within two to three hundred yards of the northwest corner of the field. Our recon activity raised awareness with the locals and several came out to investigate. We were careful and watched their movements closely. If there was something there, it's likely that they may attempt an attack to defend it. We made several attempts to get a ground team to the area but were denied because it was so far out and at the edge of the AO boundary. We persisted and continued to scout the area. Within thirty minutes of our last denial, a coalition convoy was enroute from Mosul to Tal-Afar on the same route and hit a small IED, flattening a tire on their Humvee.

Coincidently, one of the Sergeant Majors in the convoy requested additional ground support while the tire was being changed. Mad Dog Gold was then authorized to mobilize to our location. It took them about thirty minutes to get to us so we made a quick run to the FARP for fuel then made a B-line back to the location. We passed the grid location to Mad Dog and they were driving straight to it. When I say straight to it, they weren't using roads. This is what I love about the Cavalry! Their trucks were five wide and laying a dust trail that could be seen from Syria. We flew over the small mountain range that lay between FOB Sykes and the objective area, quickly catching up with Mad Dog. We flew low and fast through their trail of dust and surged forward, leading them to the objective area.

Upon arrival, Mad Dog setup a perimeter security focusing the strong side between the village and the objective. We pointed them to areas of discolored dirt and they quickly grabbed shovels and began digging as if looking for gold. We patiently waited as they dug at the first location. It only took a couple minutes and a couple feet of digging before a blue tarp was unearthed. They continued digging and found the perimeter to be approximately 15' x 15' and nearly two feet deep. In the black of night, Al-Qaeda operatives had created shallow cache points under ground and lined them with tarps to protect the IEDs placed there only hours before, most likely a staging point for use in Mosul.

We conducted a detailed recon of the remaining area, knowing that we had likely discovered one of the largest caches in our AO. We dialed in on several more areas of similar appearance and watched as Mad Dog extracted countless IEDs from the fresh dirt. As we slowly hovered near the dirt berm surrounding the field, I saw something odd, something that looked out of place. It appeared that in one area of the berm a cavity had developed. It was as if rain water had washed around something and left a small opening. With all the other findings in the area, we radioed Mad Dog and requested they check that area as well. As they dug it out, they discovered two 55 gallon barrels. Inside the barrels was an array of training videos on how to make and emplace IEDs, more homemade explosives, and several pounds of accelerants used in the manufacturing of roadside bombs.

With each cache uncovered, we kept a keen and watchful eye on the village only a few hundred yards from the field. Mad Dog continued the recon of each grid coordinate we gave them and eventually unearthed five large caches filled with a multitude of IEDs and IED manufacturing materials. Additionally, we identified four areas that had been prepared for use. The dirt was loose with tarps in place underneath.

Today was a definite win for coalition forces. We undoubtedly saved lives by taking these weapons out of enemy hands. There may be more coming tomorrow, but we'll be right here to shut 'em down. Days like this make my job worth it for me. I'm thankful for the flat tire the Sergeant Major had. Without that event, Mad Dog may have never received clearance to come out and play.

Chapter 14

The Diagnosis

15 March 2008

Today starts my two week rotation in Mosul. I threw a small bag of clothes in the back of our OH58 and departed FOB Sykes. Todd has been in Mosul for two weeks and his time is up. Shortly after landing at FOB Diamondback, I jumped out and Todd climbed in. I'm sure he was ready to get back to FOB Sykes. I tossed my single bag into a CHU and humped it to the TOC. During the morning brief, I learned that FOB Diamondback was attacked last night and several large rockets impacted the airfield. Shrapnel peppered the parking ramp and tore through five aircraft. One will require an engine replacement and the others extensive sheet metal repair.

Randy, our Squadron Standardizations Pilot, was on the ramp pre-flighting an aircraft with two of our new pilots when the rockets impacted nearby. Randy said they looked at one another then to him and said, "What do we do?" With a determined look on his face, he quickly snapped back and said, "Keep checking that logbook. They've already hit and you're okay so drive on."

His story took me back to my first day in Iraq. Thirty minutes in this country and I had rockets whistling over my head. I tried sleeping in my body armor but that did not work; I was in survival mode. That quickly changed. I'm so well adapted that it rarely fazes me now, unless the rounds are extremely close by.

The flight mission in Mosul is very different from that of our AO. Ace Troop is restricted to flying over or very near the city of Mosul. Mosul is a large and heavily populated city but flying over the same area for hours is mind numbing. This also means that it's very dangerous if you're not vigilant. Complacency can bite you at any time. We spent several hours over the city today and thankfully, it was uneventful. Time to log my flights and get some rest; tomorrow will be another long day.

21 March 2008

I've been in Mosul for a full week and learned their mission fairly well. The mission differs from our AO in that we have a much larger area to cover and we're continuously scouting. Out west, one of our missions is to keep weapons and Al-Qaeda operatives from getting into the city. I enjoy the hunt. It's rewarding to be part of a major find or capture / kill top level terrorists. Ace Troop has a good bunch of guys but I miss flying with my Troop.

The inner city is a hot zone and requires us to move very quickly at low altitudes during the day to avoid being hit by small arms fire from the ground. At 50 feet and 100 mph, it's nearly impossible to do anything other than get a glimpse of something or draw fire to yourself when zipping around the city. Having teams up continuously keeps pressure on the enemy and quills violence. It's impossible to measure the effectiveness, but people tend to stay calmer while we're overhead.

Our Intel group recently received information from informants regarding a large scale attack that is set to take place soon near Mosul. This happens from time to time. Sometimes they're real and other times they are meant to intimidate and impede our operations. I whispered a prayer for God's protection. "Keep us under your wing, Lord."

I called home today and got news that my dad had been admitted into the hospital because he was weak and short of breath. I called to check on him and he didn't sound well. I'm concerned with his health. I could do without the additional stress right now. In addition to that, Bertie said the kids are acting out and don't want to do their chores. Lord, please make us stronger through these difficult times.

23 March 2008

It's Easter Sunday and I went to breakfast with Mark, a fellow pilot from my Troop in Tal-Afar. After breakfast, we were walking back to the TOC for our mission briefing and just before arriving, there was a very large explosion that rocked the FOB. The explosion was sharp and piercing. It seemed very close so we both quickly moved to a nearby concrete barrier. It only took a few seconds to identify the origin. We looked over the city and saw large plume of black smoke a few miles west of the FOB. I couldn't believe it was that far away. It was so loud my ears were ringing. It sounded as if it were inside the wall. This must have been a very large VBIED. Mark and I were both startled to say the least.

It wasn't long before the reports began coming in. A suicide bomber drove a dump truck full of HME through a security check point of an Iraqi Army training camp and detonated it in front of their three barracks buildings. The buildings were several stories tall and constructed of concrete. The bomb was very effective and killed a multitude of new Iraqi recruits.

We flew over the city for several hours. After returning to the FOB, two more VBIEDs were detonated at the entrance of Iraqi compounds. Both were in areas we had been flying all morning. Had we continued to fly, would they have attacked? I wish I knew. After landing, I attended a barbeque the Chaplain had put together and then watched a baptism. This reminded me that today is Easter and what I had to be thankful for. "Thank you, Lord, for life and for sending your Son."

25 March 2008

VBIEDs have been the hot topic lately. During our intel briefing, we were given information on several VBIEDs to look for. We aided a ground team in finding one of the bombs. Once we had it identified, a large area was cordoned off. This particular VBIED was a small car. It was riding very low in the back, from the HME that loaded it down. The ground team kept people clear of the area while elements of the Iraqi Army fired at it. The car detonated after only a few rounds penetrated it, as if it were scripted in a movie. It was quite an explosion that rocked the east side of Mosul and left a large plume of smoke above the city. We felt the percussion in the helicopter. It felt as if the aircraft shifted sideways. Another VBIED off the streets is good news.

We've been on standby for a mission to kill a high valued target (HVT) in the city. It was cancelled just before go time. After five hours of flying, we wrapped up our mission day. I devoured a bottle of water during the mission debrief, and then called home. After chatting with Bertie, I had supper followed by a cigar to help me relax and unwind.

I called my dad to check on him and he sounded worse. The diagnosis.....cancer and it's bad this time. The doc says it's in his lungs, liver, and bladder. I keep praying for God to save his life. My mind is full of scenarios. It's not a distraction during the mission but I think of him often when I'm not in the air. The faces of death in Iraq don't seem to affect me but the thought of losing my dad makes me physically ill. I know this is in God's hands. If his time has come, I can accept that.

26 March 2008

I was crewed with the Squadron Commander today. I always enjoy flying with the SCO. Right out of the gate, we were once again on the hunt for more VBIEDs. We found several that matched our given descriptions and began calling them in to the ground teams. I spotted a dump truck that perfectly matched a description given that morning. Most VBIED dump trucks have tarps covering their load so it's impossible to see what there carrying, as this one did.

A few minutes after we sent a spot report on the truck, it began moving across the city. We made the decision to stop the vehicle until our ground units arrived. I was flying right seat and on the controls. As I flew over the truck, the SCO dropped a smoke grenade about 50 feet in front of the truck to signal the driver to stop. The driver navigated around the smoke and kept moving. We dropped a second grenade but once again the driver went around the smoke and began to move faster. It seems this driver is fleeing and nervous, I'm certain that this is one of the VBIEDs! The SCO pulls out his M-4 and asks me to put him in position for a burst of warning shots. I maneuvered the aircraft into position and he squeezed the trigger laying three rounds directly in front of the truck. The driver stopped after the warning shots were fired but hurriedly backed up, changing his direction. This guy isn't stopping!

Our ground units are en-route but are still a few minutes from our location. The truck is in an area that is sparsely populated but he is moving west into the main population. If the driver is allowed to move the truck much further and detonates the bomb, the casualties could be devastating. At this point, we are deliberating whether we should kill the driver. After further discussion and a radio call, we receive clearance of fires and proceed to setup for a precision shot with the M-4, as any other method may detonate the truck, killing us and anyone within a few blocks. I positioned the aircraft to the left flank of the truck. This position assures missed shots won't detonate the truck. He anchored his M-4 against the door frame and began dialing in his sight picture. As he called his weapon from safe to fire, I spotted our ground team rolling only a couple blocks away. "Hold your fire!" I said. The ground team is here. Let's see if we can roll this guy up. He agreed and we verbally walked the team onto his location.

We continued to provide air support and security for our ground elements. As the team approached the dump truck, the driver jumped out of the vehicle and got into a passing car. Our ground vehicles rolled up in force, five strong, and drug him out of the car. They interrogated the driver for approximately ten minutes. Part of the team searched the truck while two other soldiers rolled the tarp back. The truck was full of dirt. No explosives, no weapons, it was nothing but dirt! Why does a guy run when he's not guilty? I don't understand why but I'm sure glad I spotted our ground team and called off the shot. I feel good knowing we didn't kill an innocent man.

Not long after this incident ended, a suicide bomber drove a VBIED into an Iraqi security checkpoint and detonated it. It was quite a blast. Small trucks can hold a lot of explosives. Both vehicles burned and I never heard if any Iraqi soldiers were injured. As the day progressed, we responded to two more calls, one of which was for troops in contact, commonly referred to as a TIC.

We typically have two scout weapons teams up flying over the city, but the west side team was down for a quick lunch so we were alone in the air. A report came in that a coalition convoy was moving through the west side of the city and was attacked by small arms fire and a grenade. The location was near a mosque, which many times is like the bee's nest. They use mosques as a safe house, knowing it's against our ROE to fire on it. We will fire on a mosque if the Al-Qaeda fight us from it. The intent of the ROE is to respect their religion and churches, but there is a limit, we are allowed to defend ourselves.

To date, I've only flown the eastern side of Mosul. Historically, the western side has been more of a hot zone, according to the SCO. The area that we were going to was very near the location where one of our Lieutenants took an AK-47 round through his leg. As he programmed the grid into the navigation system, he briefed me. "We have to stay low and fast." I could hear the intent in his voice. "There's no forgiveness for mistakes." I was dialed in and my senses were peaked. He unstrapped his M-4 from the dash as we pushed into the western side of the city, looking for the gun fire. I pulled as much power as the bird would allow and we were on the objective within two minutes. As we arrived, we saw our convoy below, just down the street from a mosque. He said, "Make a wide orbit around the objective but keep it fast." The low bird in the team normally flies between 50'–100'. I pushed it lower. I dropped to about 25' over the buildings and our airspeed was nearly 100 knots. My heart was pounding. I threaded the aircraft through several antennas sticking up from the buildings below, narrowly avoiding each and keeping the aircraft as low as possible. The low altitude and high speed reduces the window of exposure to the enemy. We made two passes of the area and there was no one to be found. Sometimes I wish they would fight back so we can eliminate them. In cowardly fashion, they come out and fire a few rounds then disappear into the walls of the city. The convoy quickly evacuated their wounded personnel while we flew overhead. We attempted to lure the enemy out but they were finished playing and were never seen again.

After the incident, another team launched and we returned to the FOB to call it a day. After we shut the bird down, the SCO jumped out, crawled under the front of the aircraft and was lying on his back, looking at the chin bubble area. I walked around to his side and said, "What are you doing, sir?" He replied, "I think you hit an antenna." I laughed and said, "I didn't hit an antenna." "Are you sure?" he asked. "I'm positive! Let's go get some food." It was a busy day. After debriefing the S2 Officer, we made tracks for the chow hall.

We had a late lunch and then I dropped off some laundry. As I'm sitting in my CHU writing this, I can hear helicopters flying and bombs detonating around the city. It's someone else's fight while I sleep. Tomorrow is another day. It's crazy what I've gotten used to.

27 March 2008

It's my mom's birthday today so I called and wished her a great day. I asked to talk to my dad but he didn't feel like talking. This is a first. Dad always wants to talk to me. I'm concerned about his health and recovery. He's sounded progressively worse each time I talk to him in the past two weeks. Mom said he's not eating well and he's weak. Even though this is out of my control, I feel more helpless being here instead of at his bedside.

29-31 March 2008

Again today, mom has told me that dad is in worse shape and she isn't confident he will recover. She is overwhelmed by both his situation and trying to care for my grandmother. She's been living with them for a few months now. At the end of our conversation she said, "If you boys want to see your dad alive, you better come and do it." The hair on the back of my neck stood up. I had a very empty feeling that my dad was about to die without hearing a goodbye from me. It brought tears to my eyes.

I hung up the phone and called my sister. She wasn't sure if mom had come to terms with the fact that dad was only a few days from dying. I told her the only way I would be allowed to leave Iraq was through a Red Cross message. I told her to contact the doctor immediately and inform him of my situation. She made contact with the doctor at his home around 2200 hours stateside and relayed the message.

We were on mission early the next morning when I received the call over the radio asking me to return to the TOC. Even though I knew it was coming, it still hit me hard. The message said that my father had terminal cancer and was expected to die within two weeks. Tears ran down my face as I walked to a bench just outside the TOC and sat down. Up to now, everything I've been told was informal. Reading this official message from the doctor made the situation very real to me. I was pulled from the mission and replaced by another pilot.

The S1 shop completed my emergency leave packet and sent it to the Brigade level for approval. I called my Commander, Captain Sova, to inform him of the situation. He had one of the flight teams go to my CHU and pack my civilian clothes. The team flew my clothes to Mosul a short time later. We're required to carry a set of civilian clothes in case something like this occurs. I always pack them, never thinking I'll actually use them. Captain Difabio remained with me the remainder of the day and accompanied me to the waiting area for my flight that was scheduled to arrive around 2200 hours. I thank God and my unit for getting me out quick. Within twelve hours of notification, I had made it to Kuwait and was on a plane to DC. It takes a long time to fly half way around the world. I prayed that he would hang on until I could see him. I made it through Germany, DC, and finally to St. Louis. I arrived in St. Louis at 1100 hours on the 31st, the day before my dad's 65th birthday.

Bertie and the girls will be flying in from Seattle. She doesn't do well with traveling on her own in stressful situations so I hope she doesn't get overwhelmed. My in-laws, Bob and Marilyn, picked me up from the airport and we made the two and a half hour drive home. As I walked into his hospital room, he turned his head to see who it was. His face lit up and I could see the relief as his eyes began to tear. I think he was more relieved that I was okay and had lived through the war than he was of his own situation. He didn't look well. His face was pale and swollen. He sounded as if he worked for every breath. He smiled and we greeted each other with a hug. The whole situation touched my sister and she began to cry.

My mom was in the room and looked very tired. Although terrible circumstances, it was great to see everyone again. I recommended she get some rest and let me sit with dad for the day. Mom went home to clean up, get some rest, and care for my grandma.

Talking was difficult so we limited conversation to only the necessary stuff such as fishing memories and good times together. My sister and I spoke to the doctor while he napped. The doctor said he didn't see him living more than a few days. This upset my sister once again and it was all she could muster to hold back the tears.

1 April 2008

It's my dad's 65th birthday today. Several friends and family members have been in to visit him. It's difficult to tell but I think he likes the visitors. He's well sedated for the intense pain that he's in. My brother, Rod, is coming in from Alaska soon and will be bringing most of his family. Mom came back to the hospital so I went to her house to clean up and see my grandmother. She was glad to see me and was very worried about dad. She kept saying it wasn't right for him to die before her. She appeared much frailer than when I had last seen her, just over a year ago.

2 April 2008

I woke up early and made breakfast for all of us, even mom's dog. After breakfast, I went back to the hospital to check on dad. Mom said that dad had kept his glasses on all through the night and was very tuned to the clock and the passing of time. His kidneys had shut down over night as a result of the liver's malfunction from the cancer. Rod and Beth are scheduled to arrive tomorrow. I hope they make it in time to talk to him.

3 April 2008

Rod and Beth arrived today and we all gathered around his bed. We shared stories and dad continued to crack jokes to lighten the mood as he often did. At one point, he pointed to the wall and said, "Is that a door or a gate?" Chills covered my body and tears came to my eyes. I knew his transition was near. Later today, his body made a turn for the humanly worse but a heavenly better. He closed his eyes, slipped into unconsciousness, and continued to breathe with great labor. I hooked up my laptop and began playing some worship music. Although unable to communicate, I knew that he could hear us as tears ran down his face. As his body continued to fail, the pain became more intense. The doctor increased his dosage of morphine to keep him comfortable.

4 April 2008

My in-laws picked up my wife and daughters from the St. Louis airport around midnight and drove straight to the hospital. They arrived at 0230 hours this morning. The girls came into his room and both hugged him and told him they loved him. They were both shocked to see him in this condition. By the look on Naelyn's face, she knew he wouldn't make it. He has not been conscious for most of the day. My in-laws took the girls for the night.

At 0620, dad's breathing pattern changed and at 0630, he passed. The moment he took his last breath, he had a big grin on his face. It was as if I was watching his spirit leaving his body. One can only imagine the transition to eternity until it's our turn. I love you, dad.

6 April 2008

Dad's visitation drew hundreds of people to the funeral home. Most were people I knew but others I had never seen. This is a true testament of how many lives dad had impacted. The line extended out the door and down the sidewalk, all had come to pay their respects. Life isn't about how much money you can amass or what positions you've held. You can't take any of that with you.

7-12 April 2008

Having been deployed to combat for ten months, I cherished the family time while everyone was together. Bertie and I went out to dinner with Rod and Beth the night before I went back to Iraq. I appreciate my family and friends so much more after surviving my combat tours. Problems seem so much less significant because of what I've experienced. Life is much less stressful when you're not in a hurry. Stop and smell the roses now and then.

13 April 2008

It's Sunday morning and I'm scheduled to fly out of St. Louis this evening. I'm not feeling well this morning. I'm not sure if I'm sick or if it's the anxiety of going back into Iraq. Bertie and I attended church with her parents and I continued feeling worse throughout the day. By the time I reached the airport, I felt like I had the flu. I arrived a couple hours early so I went to the USO in the airport to lie down and hopefully recover somewhat before my flight. There was no relief. My stomach was killing me and it was time to get through security. I waited until the last minute. I joined the line at the TSA checkpoint and suddenly had a chill and knew I was going to puke. Like a criminal, I quickly ran from the checkpoint toward the first trashcan I could find. I'm surprised they didn't chase me because of how I acted. As I rounded the corner, I was beginning to choke and cough. I looked up and there was a restroom sign. Oblivious to gender, I pushed open the door and ran to the first sink and threw up profusely several times. Luckily, I had run into the men's room. I felt terrible. All I wanted was a dark room and a cold rag.

Shortly after "the purge" and splashing some cold water on my face, I felt much better. I made my way through the TSA checkpoint and on to my gate. Feeling like death, I boarded the plane. The lady next to me could tell I wasn't well and asked if I needed some ibuprofen. I graciously accepted and we began to talk. I quickly learned that she worked for the Department of Defense. She said she was a director in the office that prints and distributes all of the military flight publications. We visited throughout the flight and I eventually began feeling better. After landing in DC, we went our separate ways.

After 18 hours in the air, I landed in Kuwait. I had a lengthy bus ride to the FOB in Kuwait and waited two days to catch a ride on a C-130 into Mosul. A few hours later, my Blackhawk brothers picked me up from Mosul and took me to FOB Sykes in Tal-Afar, where I re-joined my Troop. My Commander gave me a couple days off to adjust to the time and environment change before climbing back into the cockpit. Jumping right back into the mission is exactly what I need to keep my mind off my dad's death.

Chapter 15

Pucker Factor

6 June 2008

It's been a while since I've put words to paper. I've avoided writing because my mind tends to drift back to my dad's death. Since my return, I went back to Mosul and flew with Ace Troop while Major Nicholson was on R&R. I spent many hours in the air, scouting and rounding up bad guys.

One particular mission night, I was flying left seat and the TOC tasked us with a short notice mission to provide aerial security and close combat support for a special ground team. "Blackdeath 23 this is Redcatcher Oscar." I replied, "Blackdeath 23 go." The TOC followed with, "Land on Alpha taxiway at north parking for your mission pack." We promptly returned to airfield, picked up our mission pack, and took on fuel at the FARP. We were the lead ship and I was left seat, which means I'll be the primary point of contact to the ground commander. It was the wee hours of the morning and we were supporting a mixed tactical team. The ground team consisted of Delta Force and CIA. I made a radio check in with the ground team and they were already on the roll. In short order, I had the mission data programmed into the aircraft and we conducted a quick recon of the objective area prior to their arrival. The area was a populated neighborhood but it was well past curfew and all was quiet.

For the sake of operational security, I'll call the ground commander Ghost 6. After the initial recon, I called the inbound team, "Ghost 6, Blackdeath 23, there's no movement at the objective and negative movement on the roof-tops." He replied, "Roger Blackdeath." As they moved closer to the objective, the convoy stopped a couple blocks away. A small team moved tactically to the objective residence. Most of the team remained just outside the wall of the yard, but a small group covertly penetrated the perimeter.

I could see the team messing with a vehicle in the driveway, but was unable to see exactly what they were doing. We maintained an altitude that would mask the noise of our aircraft. This mission wasn't a snatch and grab as many are. A few seconds later, they entered the residence very quietly. They were inside only three minutes or so, then quietly moved out. The team retreated to the vehicles and made their way back to FOB Diamondback. The entire mission took no more than thirty minutes. I'm not sure what their mission was that night, but I suspect they wanted to track someone. Our job was to provide aerial reconnaissance and be available for close air support if needed. I love missions like this, get in and get out, like ghosts in the night.

During my ten day stint in Mosul, we had two new Lieutenants assigned to our unit. 1LT Calvert was assigned as our new Platoon Leader. He flew with our SP, CW4 Morris, for his initial training and then was pushed to me for his mission training. After only a few flights, I am convinced that this guy is bad luck. Every time we climb into the aircraft together, it breaks! I was having emergencies I've never had.

One specific instance I recall, Lieutenant Calvert and I were flying lead on a night recon mission over the Tigris River, southeast of FOB Q-West. It was approximately 2300 hours and a zero illumination night. I was in the right seat and on the controls the majority of the time while he was on the sight, using thermal imaging to recon the river. We were looking for smuggling activity and suddenly I hear, "BING!" I look down at the MFD and see, "FUEL FILTER BYPASS." This message means that the engine fuel filter is clogged. If I was stateside, I would immediately land the aircraft in a field or on a road, as this is an urgent situation. But in Iraq, there is no way I'm landing out here. I'll take my chances. We made an immediate turn toward FOB Q-West and radioed trail, "We've got a fuel filter bypass up here and we're headed directly to Q-West." We were in a very hilly area along the Tigris River so I asked Lieutenant Calvert to bring up the RMS page on his display. The RMS page displays terrain relief on a map so I could be sure I was flying over the flattest terrain possible, just in case our engine failed and we had to make an emergency landing. Situations like these elevate a pilot's "pucker factor." At this point, my pucker factor was pretty high. We were over 30 kilometers from FOB Q-West and I was praying that the little squirrel inside the engine compartment behind me would keep running until we made it to the FOB.

I called Q-West Tower, "Q-West Tower Blackdeath 23."

Tower replied, "Q-West Tower, go ahead Blackdeath 23."

Me - "This is Blackdeath 23, we are approximately 30 k's to the south and declaring an emergency. We're inbound with a fuel filter issue and requesting an approach direct to the west ramp upon arrival."

Tower - "Blackdeath 23 copy all, you're cleared direct we have no other traffic."

I asked trail to send a BFT message to our TAC back at FOB Sykes and advise the maintenance guys to get a team spun up and ready. Like the great team members they are, they had beaten me to the punch. I flew the aircraft using a low power setting so it would demand less fuel and hopefully fly long enough to get us to the airfield. Lieutenant Calvert and I reviewed the downed aircraft procedures in case the engine were to fail before we were inside the FOB. We had calculated station time for the trail bird and knew they would have less than an hour overhead if we had to make an emergency landing. The flight back to Q-West would take approximately twenty minutes. Our airspeed was 70 knots and the flight seemed to take forever. I made progressive calls to Q-West Tower, sending updates every few minutes.

We finally made it over the wire and into the FOB. "Q-West Tower, Blackdeath 23 short final." "Roger Blackdeath, you're cleared to land west parking." As we came to a hover over the taxiway, the dust from our rotor wash so bad I could only move a few feet at a time. My visibility was limited to about twenty feet. A dust storm had blown a thick layer of dirt over the taxiway and parking ramp, causing us to nearly brown out while trying to land. We eventually made it to our parking ramp, landed, and shutdown. Lieutenant Calvert is an awesome Platoon Leader. I enjoy giving him some flack simply because he's the new Lieutenant.

7 July 2008

We've been here over a year now and will soon be going home. Patience is in short supply and tempers are flaring. I saw this happen at the end of my first deployment around the same time frame. There's an abundance of testosterone and people are getting short with one another. Everyone needs a break.

Matt and I linked up for dinner before our mission and he told me that an aircraft had gone down in Mosul. He said it was Andy and D.C. from our Troop, two very close friends of ours. My heart sank and I felt ill. We found out a short time later that both were un-injured. What a relief it was to hear that. The aircraft was a total loss so I'm thankful they made it out without injury.

I called my wife when I was able to let her know everything was okay. She was on the phone with Andy's wife. He did as most husbands would and gave her the G-rated version. The incident reminded us once again of how dangerous this mission is. It's good to hear Bertie's voice again. I'm coming home soon, baby.

Crashed OH58D-R

21 July 2008

We should be leaving this sand box in a month or so. We've slowly been packing our gear. It's a good feeling. We're beginning to see the light at the other end. I recently made the decision to decline my promotion to Chief Warrant Officer 3 and get out of the military after my commitment is up. I would stay if the Army would guarantee me a transition to Fort Rucker with the instructor pilot course but I can't get them to commit to it. Right now is a critical time in my daughters' lives and I can't afford to take the chance that I might be deployed again soon. Our unit has already told us that we'll only have a year stateside before our next deployment. The girls need me right now and they have to come before the Army. We should be going home in just over thirty days. Lord, I pray you take each of us back to our families safely.

27 July 2008

I rolled out of bed this morning and found myself staring at a pile of American flags I had brought to Iraq to fly for various people. I'm not sure why, but I felt an urge to fly them all today. I piled them all together, I believe there were seven, and took them with me to the CP. After dropping the flags off at the CP with my flight gear, I hoofed to the chow tent.

At breakfast, I linked up with Kevin, Matt, and CPT Whitney for our mission briefing. We had a fairly routine mission tasking for the day, nothing too spectacular. Both our aircraft were configured with rockets on one side and a single hellfire missile on the other. Kevin and I were in the lead ship. He was flying right seat, I was in the left. Matt was flying right seat in the trail bird and had CPT Whitney in the left.

We conducted a route recon from FOB Sykes to FOB Q-West and had nothing to report. Just before reaching FOB Q-West, we were tasked with a change of mission. Bandit Troop with 1/3 ACR out of FOB Q-West had intel on the location of an individual they wanted to take down and requested air support during the mission. Our TOC sent us the grid location, frequency, and time on target (TOT) for the mission. We had approximately 45 minutes before game time so we hit the FARP in Q-West and topped off with fuel.

ME - "Bandit this is Blackdeath 23."

BANDIT - "Blackdeath this is Bandit, go ahead"

ME - "Blackdeath 23, we're your scout weapons team today, you've got two Kiowas carrying eight rockets, two hellfire missiles, and have 90 minutes of station time over"

BANDIT - "Roger that Blackdeath 23, request that you hold to the northwest of the objective until ten mikes prior to go time"

ME - "Wilco"

Approximately twenty minutes prior to go time, we departed the FARP and moved 15 kilometers northwest of the objective. Bandit was enroute to the objective area with a platoon sized element and confirming the intel with the source. The objective area was a small village northeast of FOB Q-West. We had ten minutes to burn so I tried to pinpoint the objective house using the sight. It's difficult to do from that distance but I was able to become familiar with the area.

At the ten minute mark, we moved into the small village and continued our recon from about three to four kilometers away. Everything appeared normal. As always, the locals were curious as to why we were in the area. Truth be known, they know exactly why we're here. A few minutes passed and Bandit rolled in from the south. No one tried leaving the area but they did disappear into their houses. Bandit elements setup a perimeter just west of the objective and dismounted from their vehicles. A team of four stacked in formation and moved to the north side of the objective house. We positioned ourselves on the north side of the house for the initial phase of the raid in order to provide immediate air support if needed.

As they approached the door, we dropped to just under 100 feet off the ground and turned toward the objective house. I had my M-4 out the left door and ready. As Bandit entered the house, we flew just overhead circling to the south side of the house. We came around the west side of the house and four insurgents ran out the south door. It was a four on four all out gun fight! Bandit was only 15-20 feet from the insurgent group and I couldn't immediately get a clear shot with the M-4. Within a second or two, two of the insurgents shifted their fire to our aircraft. We were extremely close to them by this point. I estimate no more than 100 feet separated us from the insurgents. I remember thinking, there's no way they can miss us at this range and the pucker factor pegged. The gun fire was so close to us that it was activating my microphone on my helmet. I yelled at Kevin, "Taking fire, break right!" Kevin turned the aircraft hard right. Between the ground fire from Bandit and us moving in from the air, it allowed Bandit to break contact and move back into the house. As fast as we turned away, the insurgents ceased firing and I yelled, "Break left, let's get back in there!" Kevin banked hard left and we moved right back to the objective.

One of the insurgents had been hit and was slowly crawly away. The other three moved to a small structure of about ten foot square that had no roof. I heard Bandit come over the radio, "Man down, we need immediate MEDEVAC!" Bandit sent up a MEDEVAC request to Mosul and our trail aircraft coordinated the pickup LZ. Kevin and I circled around the remaining three insurgents so we could engage them with the M-4. In the process of doing so, Bandit threw a grenade over the wall and into the room they were in. It exploded before they were able to exit. Boom! They ran out the east door and into a nearby house. I could tell by the way they moved at least one of them was injured. We had them pinned down inside the house. The house was about 40 feet long and had no windows so it was difficult to tell where they were inside. We continued receiving small arms fire from the insurgents from the doorway.

I called Bandit, "Bandit, we have the three squirters pinned down inside the house on the east side of the objective." "Pull your guys out of the area and we'll engage with hellfire."

Bandit promptly replied, "Roger Blackdeath, we'll advise when all elements are clear."

While Bandit worked to move their wounded soldier to the LZ for pickup, we put a plan together that would end this thing. The MEDEVAC Blackhawk landed not long after Bandit reached the LZ on the west side of the village. Bandit quickly loaded their soldier and the Blackhawk departed for the cache in Mosul. Bandit's wounded team member took a round through the leg during the raid.

The building the insurgents were inside of was long enough that it would take both of our hellfire missiles to guarantee the kill. I called trail and suggested a welded wing formation, simultaneous engagements of the building and trail agreed. There was one doorway on the south side of the building and several obstructions on the north side so we setup to engage from the south to the north. We flew side by side toward the house and I counted down to the shot.

Trail called as we were inbound toward the target, "I do not have positive ID."

I replied, "It's the building north of the open courtyard with the large tree."

He repeated, "Negative PID."

I said, "Roger, I have PID and we're taking the shot, follow-up with rockets."

Matt replied, "Roger on the rockets."

With only one missile, I'll have to guess their location inside the building. I positioned the reticle halfway between the door opening and the end of the house and announced, "Missile in constraints." Kevin replied with "firing," pulled the trigger and the missile left the rail, whoosh! Several seconds later the missile impacted, creating a large plume of smoke, dust, and debris. Matt followed and squeezed off two rockets, both impacting near the target.

First missile shot

Both of our aircraft were at minimum fuel after the engagement. The FARP turn in Q-West would take approximately twenty five minutes. I called Bandit to let him know we would be departing for a quick splash of gas while they formulated a plan to proceed. While we were gone, the 1/3 ACR Squadron Commander, Tiger 6, ordered additional assets be sent to the objective to aid in the fight. A platoon of Iraqi soldiers arrived to assist along with Tiger 6 himself.

As quick as possible, we fueled and returned to the village. Upon our return, the ground units were preparing for a house to house search of the entire village. We slowed to our maximum endurance airspeed to afford the most station time possible.

The Iraqi troops took the lead on the house to house search and all went well for about thirty minutes. As they reached the house we had just put a missile into, they once again came under enemy fire. Fortunately, there was only one insurgent left fighting. He fired is way out of the rubble and ran south to the next courtyard. I called to the ground unit, "Bandit, we have one insurgent on foot, running south." He was heavily armed with several bandoleers of ammo and grenades. As he approached the adjacent courtyard, he jumped the wall and ran into a house about 75 meters south of the one we had just engaged. I relayed to Bandit the insurgent's new location.

Iraqi Army preparing to conduct a house to house search

I had several thoughts going through my head. Is this guy the last of them? Are there more insurgents that we're unaware of? It's easy to get fixated on one objective and then get attacked from another angle. He fired at us several times from the doorway of the house but missed each time. The only access to the house he was inside of was on the north side. There were no other doors or windows so we avoided the area directly north of the house to limit our exposure to the doorway.

Less than two minutes after the insurgent entered the house, two Iraqi soldiers decided to be heroes and jumped the courtyard wall, landing themselves inside the lion's den. They were completely exposed to be engaged. The insurgent tossed a grenade at them. One soldier jumped the east wall before it detonated, the other wasn't as fortunate. Shortly after killing the Iraqi soldier, the insurgent made a run to the adjacent house to the west and ran inside. Both of these houses shared a courtyard.

Tiger 6 wanted to isolate this guy and lock him down so he moved an armored vehicle (MRAP) to the south of the insurgent's location, in the alley right behind the house he was in. The alley was a common sewage run off area from each of the surrounding houses so the ground was soft and the MRAP sank to its axles, immobilizing it 15 feet from the house the insurgent was in.

With no warning, two Iraqi soldiers ran to the courtyard, scaled the wall, and jump inside. The insurgent guns both of them down within seconds, shooting them several times. A second group ran over the wall, again he gunned them down and they fell like rag dolls. Kevin and I both yelled in the cockpit, "Stop running in there!" as if it would help. The radio was congested with traffic but I managed to get a call to Bandit and suggested that he tell the Iraqi commander to cease further attempts and pull all of their men back so we could engage the building.

Tiger 6 came on the net and requested that we engage the house after his guys are buttoned up inside the armored MRAP. "Negative sir, it's too close to the kill zone," I replied. I could tell he was irritated. A missile perfectly placed would not harm the ground team but I wasn't willing to take the chance. Equipment and missiles malfunction. It's not worth taking the chance. My target was 15 feet from his men. He agreed to move his guys so he sent another MRAP to their location to evacuate the immobilized unit. When the second armored MRAP moved in to extract the men, it too sunk it's left rear wheel into the soft sewage up to the axle. The second MRAP was lying against the first, and both unable to move. Eventually, the soldiers ran out the back of the vehicles and moved a safe distance to the west.

Once everyone was evacuated, the LTC called, "Blackdeath 23 you are clear to engage." "Roger, we'll engage from north to south," I replied. As we moved to the north of the village to setup for the shot, I called Matt to coordinate. Matt was carrying the last hellfire missile. Captain Whitney had trouble setting up the shot and did not have positive ID of the building. We needed to make this shot quick and precise so we chose to do a remote shot. Matt would shoot the missile from his aircraft and I would guide it to the target from our bird. I quickly programmed his missile code into our aircraft.

At 2.5 kilometers away, I marked the target with the laser. Matt called, "In constraints.....firing" and pressed the fire button. At 1.8 kilometers from the target, the missile left the rail. The radio was silent as the missile screamed to the target. About fifteen seconds later, BOOM! We watched as the smoke and dust slowly cleared. The house was completely leveled. Bandit keyed up the mic, "F#@* yeah!"

Second missile and MRAP locations

Our fuel was low and we were out of missiles. I called Tiger 6 and asked if he'd rather us make a quick fuel turn in Q-West and get back on station in 20-25 minutes or fly to Mosul where we can get fuel and ammo. The trip to Mosul would take us 45 minutes. He preferred a quick turn in the FARP for fuel in Q-West so we turned southwest and pulled max power.

As we arrived in Q-West, the tower controller warned that a dust storm was quickly approaching on the south end of the airfield. We could see it only a couple miles south of us and it was moving fast. I signaled for the fuel crew to work fast, pointing at the storm. They quickly topped us off but it wasn't fast enough. The storm was already coming around us. As soon as the ground team unhooked us, we pulled pitch and flew directly north to get in front of the storm. It only took a minute or two to fly out of it. Another thirty seconds on the ground and we wouldn't have been able to see well enough to take off. The path of the storm was moving right at our mission area and would soon engulf the region.

We called the ground commander for an updated status after clearing the leading edge of the storm and filled him in on the inbound weather. He advised that they had the situation under control and thanked us for our help. After signing off, we turned northwest and made a B-line for FOB Sykes. The flight home took 40 minutes which gave me time to bring my adrenaline back to normal levels. Now I know why I had a sixth sense to take all the flags this morning.

After rearming the aircraft, we repositioned to the parking pads. As we hovered in to land, our crew chief, Dennett, met us at the pad. I thought of the irony of being in aircraft 007 (better known as balls 7) for such a great mission. We flew 7.5 hours in her and she had performed beautifully. Dennett was happy to hear of her great performance and helped as we gave her a detailed post flight, anticipating finding a few bullet holes. Miraculously, we only found a small nick in the tail rotor. I'm not sure how we didn't get hit. God is good. I pulled the stack of American flags from the aft compartment and proceeded to unwrap one. We were all tired and soaked in sweat but we took time to pose for a quick photo while holding the flag in front of our bird. What a day it's been.

We conducted a detailed debriefing with the S2 team before going to dinner. I hate seeing our soldiers get hit or killed. Today's events will play through my head for years to come. I'm already rethinking decisions. No matter the outcome, I think everyone replays it in their mind. The battle ended with four insurgents killed, but not before they killed six Iraqi soldiers and wounded four others. One American soldier was wounded, but it could have been worse. I feel like the Iraqi lives were lost in vain. Their rushing the house made no tactical sense. I'm certain these orders didn't come from Tiger 6, if they were even ordered. I expect they acted on their own or were sent in by the Iraqi commander.

14 Aug 2008

Today is Madison's birthday. Happy birthday, baby! The incoming unit is purchasing the internet and network system from our troop. We transferred the system over to their equipment today. It feels great to pack the last few items before hauling butt out of here. Morale is pretty high, to say the least.

I'm flying Major Barber today, a friend of my brother, Rod. He seems like a nice guy. Yesterday, I flew one of their newest Lieutenants and while over some hilly terrain just east of the FOB, we had a low rotor warning come on in the aircraft. My heart went into my throat and I thought I was going to have to autorotate and make an emergency landing. I quickly picked out a place to land, then looked down at my rotor RPM gauge and it showed 0%. I knew that wasn't correct so I quickly figured out that it was an instrumentation issue. The RPM gauge jumped back up to normal range so we flew the bird back to the airfield and swapped aircraft. Every time I fly with a Lieutenant, something happens! Maybe that's why I'm on edge all the time. Geez!

15 – 26 Aug 2008

Over this two week period, we completed the transfer of authority, packed, and moved aircraft back to Kuwait in preparation for going home. I drew the long straw and was able to leisurely ride to Kuwait in the back of a C-130. Unlike my first deployment, we spent only a few days in the scorching heat of Kuwait before boarding a plane for the US. Every time I'm in Kuwait I think to myself, why do people live here?

I vividly remember the departing flight to come home. As we rolled on the tarmac, it was quiet on the plane. As the wheels lifted off and we were airborne, the loud roar of clapping and celebration broke out as this was one more step to seeing our babies back home. Some slept on the ride home; I could not. I spent the majority of my time in the back of the plane, drinking coffee and visiting with the flight crew and other soldiers. We swapped stories about what we would do first. I'm sure I drank several pots of coffee on the 24 hour journey.

After many hours, we were on final approach to land in Maine. The plane was quiet until the wheels touched down. After touchdown, we broke out in celebration. It felt liberating to be back on US soil. There was quite literally a feeling of relief that moved through me as I took a deep breath. We were once again greeted by supporters as we deplaned. The support group was always there to welcome us home and say "thank you" and clap as we walked by. It brought tears to my eyes to see that level of support by people that didn't even know me. I'm proud to be a soldier and pilot in the US Army.

From Maine, we flew to McChord Air Force Base and boarded buses that took us to the airfield on Fort Lewis, only a few miles away. Welcome home signs filled the hangar. We stood in formation, scanning the crowd for our loved ones. After a few minutes, we were released to our families. My girls ran to me and gave me the never-ending hug. Tears streamed down their faces. We were finally together again. I gathered up my bags and we walked to the car. I whispered a prayer for those who had paid the ultimate price. I also walked a little bit taller that day, knowing I served my country well. "Thank you, Lord, for bringing me home."

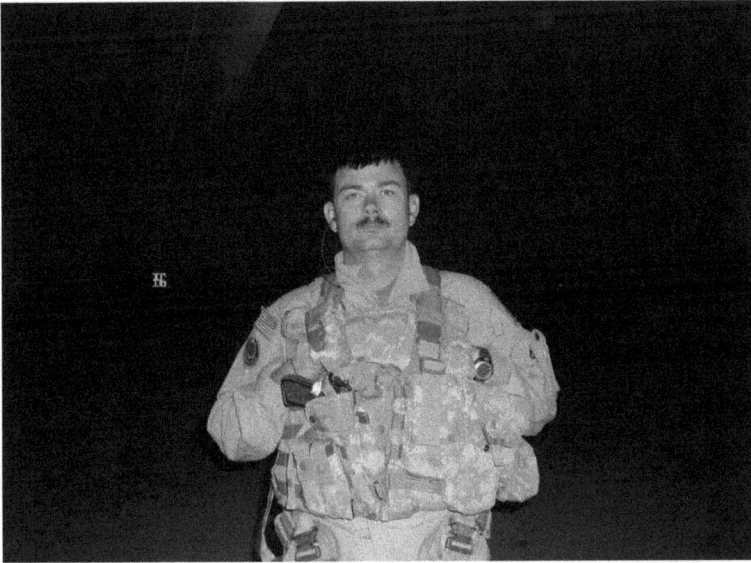

Epilogue

After my return from Operation Iraqi Freedom 07-09, I declined my promotion to CW3 and made the choice to exit the military after my commitment was complete in March 2009. My Troop Commander bugged me more than a few times to apply for Officer Candidate School and transition to the command ranks. I said, "No thanks sir. I'm enjoying the flight time." I am proud of my service to our great country and to have been able to fight alongside some of the bravest Cavalry Troopers in this nation's history.

Our reintegration into civilian life was challenging. I was hired by a helicopter EMS company and overnight my role as a helicopter pilot made an about face. I came from taking lives to saving them. With my transition came frustration, guilt, anger, and a multitude of issues. I had a shorter fuse and tolerated less. I tried not to take it out on my girls, and believe I managed it pretty well. When I felt the onset, I would separate myself from them and put up a barrier. I experienced PTSD in many ways. I wasn't the only one. I stayed in contact with many other pilots. Some were just like me, while others lost their marriages, kids, and hope for a better future.

As a family, we felt somewhat withdrawn because we had no close friends at our new home in Southeast Texas. In the military, it doesn't matter where you go, you will instantly have friends. I had a tough time dealing with the change. We barely knew anyone. Feeling withdrawn, I struggled with what to do with my free time. We were living life at top speed and now I was down to a crawl. We slowly made friends and watched our young daughters grow into beautiful young ladies. It's slowly getting back to "normal."

Acronyms

ACR – Armored Cavalry Regiment
ACS – Air Cavalry Squadron
AMPS – Aviation Mission Planning System
AQC – Aircraft Qualification Course
AO – Area of Operations
BIAP – Baghdad International Airport
COP – Combat Outpost
CP – Command Post
EOD – Explosive Ordnance Detachment
FARP – Forward Arming and Refueling Point
FOB – Forward Operating Base
HME – Homemade Explosives
HVT – High Value Target
IA – Iraqi Army
ICDC – Iraqi Civil Defense Corp
IERW – Initial Entry Rotary Wing
IED – Improvised Explosive Device
JRTC – Joint Readiness Training Center
LT – Lieutenant
LZ – Landing Zone
MMS – Mast Mounted Site
MEPS – Military Entrance Processing Station
NOE – Nap of the Earth
NVG – Night Vision Goggle
PI – Pilot
PIC - Pilot in Command
PL – Platoon Leader
PX – Post Exchange
QRF – Quick Reaction Force
R&R – Rest and Relaxation
RIP – Relief in Place
RL – Readiness Level
ROE – Rules of Engagement
RPG – Rocket Propelled Grenade
SCO – Squadron Commander
SWT – Scout Weapons Team
TAC – Tactical Command Post
TIC – Troops in Contact
TOA – Transfer of Authority
TOC – Tactical Operations Center
VBIED – Vehicle Born Improvised Explosive Device
WOCS – Warrant Officer Candidate School

www.ingramcontent.com/pod-product-compliance
Lightning Source LLC
Chambersburg PA
CBHW070839100426
42813CB00003B/675